SO-ADF-202

Smart Mice,
Not-So-Smart People

ALSO BY ARTHUR L. CAPLAN

Am I My Brother's Keeper?: The Ethical Frontiers of Biomedicine

Due Consideration: Controversy in the Age of Medical Miracles

If I Were a Rich Man Could I Buy a Pancreas?:
And Other Essays on the Ethics of Health Care

Moral Matters: Ethical Issues in Medicine and the Life Sciences

The Case of Terri Schiavo: Ethics at the End of Life
(coedited with James J. McCartney and Dominic A. Sisti)

Darwin, Marx and Freud: Their Influence on Moral Theory
(coedited with Bruce Jennings)

The Ethics of Organ Transplants: The Current Debate
(Contemporary Issues
(coedited with Daniel H. Coelho)

The Human Cloning Debate
(coedited with Glenn McGee)

Smart Mice, Not-So-Smart People

An Interesting and Amusing Guide to Bioethics

Arthur L. Caplan

ROWMAN & LITTLEFIELD PUBLISHERS, INC.
Lanham • Boulder • New York • Toronto • Plymouth, UK

ROWMAN & LITTLEFIELD PUBLISHERS, INC.

Published in the United States of America
by Rowman & Littlefield Publishers, Inc.
A wholly owned subsidiary of The Rowman & Littlefield Publishing Group,
Inc.
4501 Forbes Boulevard, Suite 200, Lanham, Maryland 20706
www.rowmanlittlefield.com

Estover Road
Plymouth PL6 7PY
United Kingdom

British Library Cataloguing in Publication Information Available

Library of Congress Cataloging-in-Publication Data

Caplan, Arthur L.
    Smart mice, not-so-smart people : an interesting and amusing guide
to bioethics / Arthur L. Caplan.
        p. cm.
    ISBN-13: 978-0-7425-4171-9 (cloth : alk. paper)
    ISBN-10: 0-7425-4171-1 (cloth : alk. paper)
    1. Medical ethics. I. Title.
    R724.C344 2007
    174.2—dc22   2006014275

Printed in the United States of America

Contents

PART IV
Engineering Plants, Microbes, and Animals

PART V
Experimentation Ethics

PART VI
Health Reform

PART VII
Human Cloning and Stem Cell Research

PART VIII
Mapping Ourselves

PART IX
Reproduction

PART X
The State of Science in the United States

PART XI
Donation and Transplantation of Organs

Is America Going to Hell?

THE PAST FEW years could easily leave you with the distinct impression that America is going straight to hell. We seem to be losing our moral minds. Scandals, hypocrisy, duplicity, and vice are the daily fare of the media. What is going on?

Hold on, you say. Things have been worse. Remember the sense of ethical angst that engulfed the nation when a former president was revealed to have dallied with an intern in the White House?

Admittedly, that was a dark time. President Clinton was rightly condemned by a chorus of conservative congressional critics for his immorality. But, some of those critics were quickly outed as having had an illicit sexual dalliance or two of their own. The bottom line on Monicagate was that we all understand that those involved were politicians, and Americans really don't expect much in the way of morals from that crowd.

These days things really are worse. Those from whom we do expect virtue are letting us down.

Our most prestigious news organization, the *New York Times*, got enmeshed in an ethics scandal of unprecedented proportion. A young reporter filled the front pages of the paper with errors, lies, and stolen material for months, maybe years. The coaches of our most prestigious college sports (football and basketball)—men who are supposed to be icons of morality in states such as Alabama and Iowa—confessed to such indiscretions as spending time with strippers, drinking themselves to oblivion, and trying to maul coeds at campus parties. And our athletes themselves do no better, be it cheating with performance-enhancing drugs or simply engaging in incredibly irresponsible behavior both in and out of uniform.

They are joined in the ranks of the morally bereft by a bevy of titans of American industry. Their ethical shenanigans have created a situation in which a large number of prominent companies are bankrupt with their leaders indicted. And millions of Americans are finding out that the pensions and benefits they were promised by their bosses are nowhere to be had.

Even the current president, who swore he would restore morality to the Oval Office, is in deep ethical trouble. We went to Iraq for the express purpose of ridding the world of a dictator and his weapons of mass destruction, but the latter proved hard to find. And an administration that promised it would not lie is watching some highly placed officials face the wrath of special prosecutors and grand juries.

What about our ethicists themselves? Consider the fate of those who call themselves our moral guides.

William Bennett styles himself as something of a national moralist. Bennett built his career on wagging his finger at sin and encouraging young kids to read snippets of wisdom about how to live that he thoughtfully assembled and sold for great profit in various collections. But, Bill turns out to have a huge gambling jones. The casinos know him so well that he is referred to as a "whale"—good for millions of dollars of action. Our national icon of moral priggishness can't stay away from Vegas. An aggressive effort to rehabilitate him in the public eye doesn't change the shame or irony of it all.

The same grim assessment can be offered of Pat Robertson, another self-declared moralist, who spent a good part of last year defending his un-Christian call to murder the head of state of Venezuela, Hugo Chavez. And of the Reverend Franklin Graham, Billy Graham's son, who heartlessly said that the spate of hurricanes that pounded the Gulf area of the United States were reflective of God's wrath for sin in New Orleans. Either God has bad aim, or he cannot distinguish the out-of-town gamblers and drinkers from the local population.

So what are we to conclude? Is ethics just another name to describe the values that the rich and powerful support? Is there a hypocrite lurking just behind every moral pronouncement? I don't think so.

What America needs is not a renewed effort to instill morality into the hearts and minds of its citizens. Knowing what is right is no guarantee of doing the right thing. What we need is a recognition that for ethics to work in the real world, leaders must take their ethics very seriously.

People need not just to be told to do the right thing but rewarded and promoted when they do.

Consider the case of dealing with unethical behavior on the part of children.

Ethics and kids don't seem to mix. I do not mean that kids cannot make ethical judgments or that issues of right and wrong do not command the attention of all kids from nursery school through high school. Anyone who has talked to kindergarteners about their views of the kid who failed to share knows that even very young children have a keen sense of injustice.

What I mean in saying that ethics and kids don't mix is that there are a lot of kids who behave in unethical ways. Surveys of college students show that more than two-thirds admit to having engaged in serious cheating in high school. Rates of stealing, drug use, violence, and non-consensual sexual contact are depressingly high among adolescents and even younger children.

There is a lot of bad behavior out there. Does this mean that parents should be beating themselves up over their failure to teach ethics to their children? Are parents failing, as many conservative pundits suggest, or is something else going on? I think something else is going on—we are confused about what to expect when ethics is taught.

I have one son, Zach. He went to Germantown Friends, a school with deep Quaker roots in Philadelphia. Trust me, Quaker schools yield to no others when it comes to caring about ethics. And I suspect that having a dad who is involved in ethics on a daily basis has made ethics a rather central issue in my son's upbringing as well. So, if there are kids out there who got more in the way of ethics at school and at home, I am not sure who they are!

That said, Zach, like any young person, has his share of ethical flaws and weaknesses. He is sometimes willing to lie to get out of a tough situation. He was known to skimp on a homework assignment or two in high school. Once he took credit for work he really did not do. He was not truthful about using alcohol to excess.

So what is going on? Did his dad, a full-time bioethicist, fail him with respect to moral instruction? Should the Quakers give up their teaching efforts if one of their best students engages in unethical behavior? I don't think so.

If you talked with Zach, you would quickly realize that he thinks hard about his conduct, his choices, and how others behave toward him. If he sometimes skirts out on the edge of immorality, it would be hard to argue that he does so simply because he did not know any better or lacked the ability to think through what he was doing.

Zach, like any moral agent (including his father) is tempted by all of the short-term rewards that bad behavior can bring. As some wag long ago noted, if sin wasn't so much fun, we would not engage in so much of it. No one ever said ethics was easy. And that is the point.

The test of an "ethical" child is not simply the child's behavior.

What parents need to understand is something that ethicists from Socrates and Jesus and Aquinas to Martin Luther King and Gandhi have long understood: always acting ethically is hard. That is no less true of us than it is of our children. Lapses and failures are consistent with good ethical teaching.

If we measure parental success in teaching about ethics simply by expecting perfection, we will always be disappointed. Few people, adults or kids, always behave ethically. Those who do are known as saints. For the rest of us, ethical conduct is an aspiration but not always a reality.

What a parent needs to do is to teach their kids enough about ethics to know right from wrong and then to strive to use that knowledge in a world where it is sometimes hard to do so. Kids need the tools to feel remorse when they err, to correct self-indulgence, to feel guilt about doing wrong, and to retrospectively realize that what looked like a good decision yesterday was in actuality a lousy one when seen in the light of ethical reflection today.

If parents or society expect perfection from those they teach ethics to, they will be disappointed. If instead they create the basis for remorse, reflection, self-correction, and a striving to do better, then they will have done a great job transmitting the tools of ethical character and conduct. There is still too much immorality out there to say that we are doing enough to teach our kids. But what should make us truly worried is when we do not hear much in the way of apology, self-criticism, or remorse on the part of our children, our civic and business leaders, and our moral leaders.

Ethics will only flourish when checks and balances are in place. If an organization lets an individual plagiarize or cook the books either because no one asks or no one tells, then they are as much to blame as the person himself for ethical failure. At the end of the day, ethics is as much a social activity as a matter of individual choice. If it is failing in our nation, then we have to bear some of the blame ourselves, but we also have to blame a community that is long on talk but short on accountability.

So what needs to be done? We need to be sure that editors and publishers are held accountable along with reporters who do wrong. We need to tell boosters and alums that their gifts are no longer welcome if those they support bring discredit to their university. We need to hold to account the stockholders and investors who put up with the unethical practices that have made a joke out of the notion of business ethics. We need to make sure that no one gets a golden parachute from a company where the retirement benefits have been depleted. We need to be sure that those who lead our military training institutions are held accountable for the behavior of those in charge.

We need a political reckoning for politicians who say one thing but then deliver nothing. And we must not believe that the only thing that matters in morality is what each individual chooses to do. What matters just as much is what happens in the wake of immorality both in the heart of the person and in the behavior of the community.

Part I

General Interest

Duty versus Conscience

AMERICANS ARE very concerned about the privacy of their personal medical information. They have lots of reasons to be. Recent battles over the right of pharmacists to refuse to fill legitimate and medically necessary prescriptions from physicians based on the grounds that to do so might violate the pharmacist's personal ethical beliefs are putting your right to privacy in grave peril.

Losing privacy when it comes to medical matters cannot only cause great personal anguish; it can sometimes cost you your job or your health insurance. We have all heard the stories about the drugstore chains that used personal prescription information to fuel targeted advertising campaigns to an individual's house. Imagine what you might feel when the ad for an erectile dysfunction drug is seen by your daughter or the ad for a cheaper, generic form of a drug that discourages people from binge drinking is found by your grandfather on the kitchen table. If that is not bad enough, what about those cases in which disgruntled health care workers sent lists of those positive for HIV or AIDS to newspapers or posted them directly on the Internet? And how important do you think your privacy might be if you worked for a company that self-insures but has time and again fired workers as soon as they discover, through leaks from the medical office to the human relations department, that they or their children have a potentially expensive medical problem?

Privacy was being so frequently breached that ten years ago the public pressured Congress to do something. In response, Congress enacted and the president signed the Health Insurance Portability and Accountability Act, which strictly regulates how personal medical information can be used and disclosed.

There is nothing wrong with wanting your medical history kept private. Many of us, for a variety of reasons, want to tightly control who can learn about the state of our health. We simply do not want our boss or the nosy neighbor down the street to know what is in our medicine cabinet or why we had to spend time in a hospital or clinic last month.

But we often forget why privacy is really important. Health care cannot work without it. As doctors have known since the days of Hippocrates, without the promise of privacy, there is little chance that people will be frank about the state of their mental health, the frequency of their drug and alcohol use, or the state of their sex lives. They will not admit that they have been raped or sexually or physically abused. If patients think the information will become public, then the fact that they once tried to inject themselves with heroin or can only stay awake at work by taking amphetamines will probably never be heard by the doctor. Medicine cannot work without patient honesty, and honesty requires privacy to exist.

That is why the claims of some pharmacists that their moral conscience prohibits them from filling prescriptions for birth control, RU 486, emergency contraception, and pain-relieving narcotic drugs for the terminally ill are so troubling. If you cannot get the medical treatment you need because a third party refuses to give it to you, then your privacy is jeopardized. Not only do you need to convince a doctor to help you deal with a difficult problem, but you must also convince a pharmacist. Worse, if the pharmacist makes a public statement to others about why personal values will not allow him or her to fill your prescription, then your privacy is headed right out the window.

Do pharmacists have a right of conscience not to involve themselves in actions that they feel compromises their moral integrity? If a pharmacist works for a retail chain, hospital, or nursing home, then the pharmacist's right of conscience must yield to the expectations of the institution about the nature of the job. In the army, you need to accept the reality of carrying a gun. If you work as a veterinarian, then the day will come when you will have to euthanize an animal. In health care, if you cannot engage in all the legal duties required by your job, then you should not be working in that job. So at the CVS, Eckerd, Walgreen, Rite Aid, or Duane Reade, conscience must yield to privacy whenever there is no one else available to fill a legitimate prescription.

Even pharmacists who work independently and own their own stores face limits on their right to conscience. Independent pharmacists can say no, but they are duty bound to find, quickly and quietly, another pharmacist nearby or on the Internet who can fulfill the legitimate medical orders of a physician. Pharmacy is supposed to work as a supplement to medicine, not an obstacle.

Conscience is an important right of every health care professional. But duty is more important, and the duty of every pharmacist is to ensure the privacy and thus the efficacy of the physician–patient relationship. The only way to do that is to ensure that what the doctor believes ought to be done gets done as long as it is consistent with the law. Americans should not allow their hard-won right to medical privacy to be stolen in the parking lot of their local drugstores.

The Ethics of Brain Imaging

I N RECENT YEARS, it has become possible to correlate a surprising number of different personality attributes, attitudes, and disease states with distinctive patterns of brain activity using new neuroimaging technologies. For example, some forms of schizophrenia, attentiveness, and lying can be identified from NMR scans. Scientists report patterns of brain responses to emotionally pleasant or unpleasant pictures. So it is possible to use certain types of neuroimaging to detect anxiety or distress. Even subtle prejudice has been detected through association with small changes in imaged brain responses to photographs of persons of different races and ethnic groups.

Correlations between psychological variables and imaged brain activity are not yet accurate enough to make neuroimaging a useful tool for population screening or individual assessment. However, scans may be sufficiently distinctive that they do offer useful probabilistic information. Those in police and security agencies seeking to detect criminal intent or to detain those who may pose a threat of terrorism may be willing to utilize information that is imperfect in lieu of anything else. There are also other third parties, ranging from parole boards to those involved in admitting persons to military service, that have expressed active interest in current neuroimaging capabilities. Given this interest, obvious and important ethical questions must be addressed in response to new scanning technologies.

Functional brain imaging has the potential to breach the privacy of a person's own thoughts. Preserving personal privacy is therefore a crucial moral imperative as neuroimaging evolves. It is not clear at present that no testing may be undertaken without the knowledge and consent of the person being tested. Nor are there any rules governing counseling for neuroimaging or who can keep and utilize stored images and databases. There is also no social consensus on the ways in which neurological evidence can be required or mandated by any employer, insurer, or any other third party.

It is not too late to protect individual rights in the emerging age of neuroimaging. But it is surely time to begin such a conversation.

Has Direct-to-Consumer Advertising Gone Too Far?

"**IS IT INEVITABLE** that I will get breast cancer?" asks a young woman in a TV ad sponsored by Myriad Genetics, a Salt Lake City biotech company, which aired in Atlanta and Denver. Despite women's fears of breast cancer, this kind of advertising is not a good thing.

Myriad was trying to peddle the company's test for breast cancer. Years ago Myriad got a patent on a test that detects some forms of genetic predisposition to breast cancer. Now the company is looking to make a lot of money by encouraging women to come in and get their test, which costs close to $3,000 to perform.

The problem with Myriad's ad campaign is that while breast cancer is, sadly, a very real danger, most women are not carriers of the gene that puts them at high risk of getting this disease. Breast cancer is as much the result of lifestyle—smoking, poor nutrition, and exposure to toxic substances—as it is genes. Experts say that only one in four hundred women on average will test positive for the genes Myriad can test for. Family history, not general population screening, is the best way to find out who is at risk.

Even more troubling is the fact that those who get tested put themselves at a different type of risk—economic. There are employers and insurers who may want to avoid someone who has had a genetic test for breast cancer, and the confidentiality of testing cannot be assured.

Direct-to-consumer advertising has its advantages. But scaring a lot of women into seeking out a test few of them need is not one of them.

Ethical Lessons from the Flu Bug

NEARLY AS QUICKLY as it started, the great flu vaccine shortage in the United States came to an end. By early January, many cities, counties, states, and other governmental agencies were easing restrictions on who could get a flu shot. Alternatives to the flu vaccine such as flu mist, which many people thought would fall into short supply as Americans desperately sought alternatives, were available in ample supply.

So what happened with respect to the great flu shot scare of 2004? And what lessons can be learned from the responses of public officials, industry, health care providers, and others who played key roles in deciding what to do when America faced what was initially seen as a public health crisis?

As we all know, there is not enough flu vaccine to go around. This is not the first such shortage to hit the United States—there was not enough flu vaccine last year—but this year is much worse. A huge portion of the supply disappeared when British regulators found that flu vaccine made at a plant operated by Chiron Corp. in Liverpool, England, had been contaminated with bacteria. How this mistake happened, no one seems to know—or at least no one is saying. What is clear is that forty-eight million doses of flu vaccine that were due to come to the United States did not make it to our clinics, hospitals, and doctors' offices for distribution.

This situation poses a direct ethical challenge to each and every one of us. Are you going to try to wangle your way to a flu shot even if you are not in one of the high-risk groups? Are you going to try to shove Katz out of line?

I hope the answer to that question is no. It is ethically inexcusable to put someone else's life at risk by using a medical resource that you do not need. It is very clear what the right thing to do in this situation is, but will Americans do the right thing? Maybe not. Consider these examples:

- At Louisiana State University, any student who wants a flu shot is apparently getting one. Students receiving shots at the student health center are not being screened to be sure that they're in one of the high-risk categories.
- In Pennsylvania and Colorado, jerks have broken into physicians' offices and stolen flu shots.
- Hospitals across the country are getting offers from greedy wholesalers eager to supply them with flu shots for $800 when a shot should cost about $10.
- Some doctors are vaccinating their families, friends, and longtime patients, even if they're not in high-risk categories. Meanwhile, dozens of nursing homes say their residents cannot get any shots.
- Some chain stores where many people go to get their annual flu shots are giving them out, no questions asked.
- The Internet is full of quacks and creeps offering alternative "treatments" for the flu.

It's Time for the Government to Act

If we can't guarantee that each of us will do the right thing, then maybe the government should take steps to make sure we do. So far, about the only noise from Washington concerning the vaccine emergency is that President Bush and other administration officials have been in the highly prized and closely contested state of Florida promising to supply older residents there with flu shots.

The Bush administration has also suggested Canada as one possible source for extra vaccine. But the degree of hypocrisy in turning to the north for vaccine—when the president and his health officials have been bellowing for years that drugs from Canada are not safe—is impressive. And it is not clear that our neighbors over the border will come through.

While vaccine maker Aventis Pasteur has promised to supply an additional 2.6 million flu shots by January, these doses may be too little too late. The flu season usually peaks in January, and most people should receive their shots in October or November in order to build up immunity before the worst of the season strikes.

Preventing a Public Health Disaster

We are facing a potential public health disaster, and bolder steps are needed. The federal government should declare a national emergency, and so should state officials. More than thirty thousand people die each year from the flu—and that is when we have a lot more vaccine to go around. This season could be far worse than usual in terms of deaths.

First, the government needs to take control of all flu vaccine supplies. It needs to ensure that doctors and nurses give these shots only to people, like Katz, who need them the most. The government also needs to levy stiff fines against any health care institution that knowingly gives vaccine to any patient not in a high-risk group. And it needs to lower the regulatory barriers that keep vaccines used in Canada, Britain, France, Taiwan, and Japan from entering this country.

I would like to think that when faced with a moral crunch, most Americans do the right thing. And most of us have done so in the face of this crisis. But it does not take a lot of people conniving to get to the front of the flu shot line to put other people's lives at risk.

C'mon, Washington. Let's go, governors. Make it a crime to try to shove old people, babies, and the chronically ill out of our lifeboat. Don't make Katz and others like him beg for their lives.

The Colonel
Kicks the Habit

IN THE SPIRIT of "no good deed should go unpunished," let us reflect together on the oddness of the announcement by Yum Brands Incorporated (no, I did not make that up) that they plan to make their KFC and Pizza Hut restaurants smoke-free. The corporate titans at Yum said "No Smoking" signs are going up beginning next week at the 1,200 KFC and 1,675 Pizza Hut restaurants all across the country that they own outright. It will be up to the other 6,000 outlets owned by franchisees to decide if their customers can no longer top off their extra-crispy, Pepsi, and biscuits with a smoke.

Now, I am not a critic of making restaurants smoke-free. I have no sympathy for the standard libertarian line that goes "If you don't want to eat in a restaurant where smokers congregate, eat at home." All restaurants should be smoke-free. Not just because inhaling secondhand smoke is known to reduce your odds in the long run of coming back for seconds, but mainly because smoke stinks. It ruins the taste of food.

Restaurants are places to eat. Smoking has no more place in a restaurant than does burning your trash. Smoking can be done outside. Your freedom ends at the end of my fork. There is no ethical case, none, for allowing smoking in restaurants. (Satisfied now, all you antismoking zealots? I have been assimilated.)

Nor am I a frothing critic of fast food. I like (as my bathroom scale will confirm) all kinds of food—immobile, slow, accelerating, and fast. Fast food has its place, and while I cannot stand the "pizza" at Pizza Hut (just had to get that in), the greasy, cholesterol-infused offerings of KFC have their artery-clogging charms.

No, my problem with Yum taking the high road when it comes to demon tobacco is that this is a company that took the low road in a big way when it came to health. They took the "fried" out of Kentucky Fried Chicken. They transformed themselves into KFC precisely because they were worried about whether you would buy their products if they kept telling you they were fried!

Heck, these corporate suits even got rid of their chief spokesperson—Colonel Sanders. He seemed to remind people that what they were eating might not be the most nutritious foodstuffs to be found on our fruited plain. He was sent to oblivion faster than the tabloid magazines and TV shows have managed to get rid of Britney Spears.

I don't want to be mean to the corporate masters of the universe at Yum, but as we say to one another where I work—in between bites of our Original Recipe—"Physician, heal thyself." It is hard to lead the charge to improving public health when you are slipping on big mounds of cheese, baked Cheetos, and honey barbecue wings.

Shame on Jeb Bush

SHAME ON FLORIDA'S governor Jeb Bush. Most Americans would agree that when a woman who has been raped becomes pregnant, or if she might die as a result of trying to give birth, termination of her pregnancy should be an option. Not Jeb. The governor of Florida is playing hardball abortion politics with a severely disabled woman's life when he should only be worried about her best interests.

Sometime last January, a twenty-two-year old woman living in a group home in Orlando was raped. The woman is now five months pregnant. She is severely mentally retarded. Experts say she has the cognitive and emotional capacity of a one-year-old child.

In addition to her severe mental retardation, she suffers from cerebral palsy and autism and is prone to violent seizures. These conditions make having a baby a very dangerous proposition for this unfortunate young woman. She could die if she tries to deliver the baby.

Once they learned of the rape, Florida social services moved to appoint a guardian for her. In this situation, a guardian is required to grant permission for the woman to receive a thorough medical evaluation. That's when Jeb jumped in.

Despite the fact that the woman is severely disabled, has been raped, and might die if allowed to give birth, Bush felt that the appointment of a guardian for her was not appropriate. Instead, he moved to stop the appointment of the woman's guardian until a second guardian could first be appointed—specifically, a guardian for the fetus.

A Florida judge has refused to rule on Bush's action based on technical legal grounds. As a result, we now have a situation in which a young woman, who has so many medical problems that she may not be able to survive giving birth, must go without proper medical attention. Even if she does not die in the birthing process, given her emotional and psychological problem, the experience could emotionally devastate her.

Someone must be appointed to protect her. But the governor, who's busy playing abortion politics, is not letting that happen.

13

The courts might help, but the courts in Florida move slowly. A pregnancy does not. If the fetus gets much older than twenty-four weeks, the standard age of viability, then it will be far more difficult for doctors to terminate the pregnancy. And, if the medical facts show the woman must have an abortion in order to prevent her own death, it will be a far more dangerous procedure.

In short, Bush is not thinking about the disabled woman. His focus is solely on the fetus. In this regard, he is simply wrong. His ethical focus ought to be on the young woman.

A victim of rape who might die as a result of childbirth should not be forced to carry a pregnancy to term. Furthermore, a disabled woman who needs a guardian in order to get proper medical attention should have a guardian—yesterday!

Someone needs to determine the facts in this case and decide what is indeed in this woman's best interest. That person ought not be a governor who wants to play politics with her life.

If Bush has the time on his hands to personally get involved in this case, he should first appoint a guardian for the woman—and then figure out what he can to ensure that other severely disabled women in his state are not at risk of rape. Perhaps the governor could take steps to make sure that young women like this have access to birth control or, at the very least, that they have adequate protection against sexual predators.

So far all Bush has done is put a helpless woman's life in grave danger.

Stark Raving Madness

LET'S BE CLEAR about one thing: there is no question, none, that Russell Weston is stark raving mad. In July 1998, Weston killed two police officers in the U.S. Capitol. When asked why he had driven across the country to brutally execute the two policemen, he told government psychiatrists that he did so to prevent the spread of a disease.

He said "Black Heva" is carried by the victims of cannibals who are taking over the United States. The slain officers, Jacob Chestnut and John Gibson, were two of the cannibals, Weston insisted. But, in his mind, their deaths were not a problem since as soon as he finished communicating with "the Ruby satellite system," time would be reversed, and they would come back to life.

Get the picture? Weston was obviously incompetent when, without provocation, he ran through a metal detector and shot the guards dead.

So if Weston is so patently insane, then what is he doing in an isolation cell at the Federal Correctional Institution in Butner, North Carolina, hiding under his blanket raving incoherently? Why is he not in a psychiatric hospital getting a lot of medication?

The answer is that the lawyers defending Weston know that he faces the death penalty if he is convicted of murdering the two guards. Since there is no question that he killed them, his only defense is insanity.

Yet without his lawyers' consent, Weston cannot be given any type of antipsychotic medication. They fear that with medication, Weston might appear sane and thus get the death penalty. So for two years, Weston has sat in complete isolation in a bare prison cell without any medicine.

What is going on here is nothing short of barbaric. Weston's lawyers believe that the only sure way to keep him from being executed is to keep him insane. This position is nuts.

Weston's lawyers do him no favor by keeping him alive but insane under conditions that are gruesomely cruel. If Weston were a cat or dog kept in the state he is in, protests would be mounted posthaste to help him.

Worse still, if Weston cannot be defended and found worthy of a permanent stay in a mental hospital by reason of insanity, then there is no

15

reason to pay lip service to the insanity defense at all. Nor is there any reason to have a trial. Weston should simply be killed since his role in two murders is not in dispute.

Lest there be any doubt about his mental status, consider that Weston had presented himself at St. Peter's Community Hospital in Helena, Montana, before he reached Washington, D.C., asking for removal of a chip in his head that kept him in communication with representatives of Russia. During an extremely bizarre visit to CIA headquarters in 1996, he was videotaped spewing crazy stories and rambling incoherently. His family made him move away years ago after struggling with his mental illness without success.

His lawyers know all this. Yet they seem to believe that the only way anyone will think him insane at the time he killed the two guards is if he is kept in an isolated prison cell.

Weston should receive psychiatric treatment immediately whether he gives his permission or not. Then he should stand trial. His lawyers should present his life's history, plead the insanity defense, and sit down. Any judge or jury that convicts this man as culpably guilty of murder should be summarily dismissed. Weston should then be taken to the nearest prison medical facility for the rest of his days.

And all of us should think very hard about how it is that the only way one of our most vulnerable citizens could get anyone's attention in America was to kill two guards in the halls of Congress.

Part II

End of Life

Million Dollar Baby

TWO WARNINGS about this column: First warning—**Extreme PC Danger!** If your politically correct sensibilities are easily offended, read no further. This column crosses into a topic that really gets people's ire up—attitudes about those with severe disabilities. Second warning—**Ending of a Movie Revealed!** If you don't want to know the ending of the movie *Million Dollar Baby*, read no further.

Million Dollar Baby, starring Clint Eastwood, Hilary Swank, and Morgan Freeman, has already won a variety of honors. It has received seven Academy Award nominations, including best movie. It should. It is a fine movie that offers a touching view of an evolving relationship between an aging fight manager and trainer, Frankie Dunn, played to the hilt as a craggy, old-school grouch by Eastwood, and a late-to-the fight game, white trash girl who has seized on boxing as her ticket to self-satisfaction, Maggie Fitzgerald. She is played with great poignancy by Hilary Swank.

What has made the movie controversial is the turn it takes at the end. Maggie goes after the title against a tough, veteran boxer who is willing to do anything to win. The dirty older pro cheap-shots Maggie after the bell rings, and she falls, breaking her neck.

Paralyzed from the neck down, bed bound, and permanently on a ventilator that pushes air through a tube in her throat, Maggie decides that her life is no longer worth living. She asks Frankie, who has become her closest friend and a surrogate father, if he will kill her. Frankie resists. His conscience and his Catholicism make mercy killing ethically off-limits.

Eventually, though, he decides that he owes it to Maggie to help her die. He brings a massive dose of adrenaline to the hospital where Maggie lives, disconnects her ventilator, shuts off the machine's warning alarm, and injects her with the drug. She dies moments later. He heads out of town, presumably to escape any legal action that might ensue.

The very idea that a severely disabled person might decide that her life is not worth living has driven various folks in the disability advocacy movement, a few highly visible figures on the right-wing talk show

19

circuit, and some in prolife circles into a frenzy. Marcie Roth, director of the National Spinal Cord Injury Association, said she hates the film's ending because so many people still think that "having a spinal-cord injury is a fate worse than death." "Unfortunately," she told the Associated Press, "the movie is saying death is better than disability." Conservative Rush Limbaugh, who has struggled with addiction to painkillers without feeling the need to defend this moral lapse, could not contain himself from ranting about how bad this movie is. He bashed the movie as the product of left-wing, secular thinking. Michael Medved, another right-winger who seems to think Hollywood exists only to find ever more sleazy ways to corrupt the morals of the American public—but who manages to see every movie these moral slackers release—has seethed about *Million Dollar Baby* as well. And the websites and blogs of prolife groups around the country are teeming with gripes that Hollywood dare promote a proeuthanasia movie for so many Oscars.

I rather doubt there is a conspiracy in Hollywood to promote euthanasia. Granted, recent cinematic offerings involving Ben Affleck, Jennifer Garner, and Jennifer Lopez do make the idea of mercy killing attractive, but this would leave too many agents in Hollywood unemployed. But, seriously, can even Rush Limbaugh expect you to buy the idea that Clint Eastwood is part of some vast liberal conspiracy to get Americans down on disability and all hopped up about euthanasia?

Let's talk some frank talk about the disability movement. It is not fair to lump together all forms of disability. Not all disabilities are the same. A person who is deaf or who has lost a limb or who must move about with a wheelchair is not equivalent to the challenge someone faces in deciding whether or not to continue life-supporting medical treatment paralyzed, incontinent of bowel and bladder, breathing on a ventilator, and "eating" through a tube implanted in their stomach. Movie audiences can distinguish different levels of disability. Can those in the disability movement stop saying that wondering whether you would want to live trapped in a bed forever, capable only of moving your eyeballs, makes you undervalue life with a disability?

Nor is there only one right answer to severe disability. Christopher Reeve, who I know contemplated suicide many times, decided to live on despite his paralysis. Some people, stricken with ALS (Lou Gehrig's disease), devastating spinal cord injuries, end-stage cystic fibrosis, and massive strokes that leave their bodies useless, do not. Dealing with quality-of-life decisions about these devastating types of disability in a thoughtful way in a movie, book, or theatrical presentation is not selling any kind of value message about disability or, for that matter, promoting a particular ideological agenda. It is using the medium to explore some very tough ethical questions. And that is a very good use of cinema, one that ought to be celebrated, not denigrated.

So if it is not beyond the ethical pale to ask questions about continuing to live with a devastating disability and what a person of faith and conscience would do when confronted with a request for help in dying from a person in such a state, then how well does the movie actually do in depicting these issues? There is one glaring problem. But it is not what the PC police or simple-minded right-wing talk show hosts think it is.

Maggie asks her trainer to help her die. And he does. But in the real world, she would not make such a request of her friend. She would ask her doctor and nurses to shut off her ventilator. She would ask to be allowed to die, not killed. The doctors would have to decide what to do. The ethics of allowing to die are thorny enough in their own right, but they are not the ethics of assisted suicide.

Requests to stop life-prolonging treatments are made and honored every day in American hospitals and nursing homes. Jehovah's Witnesses say no to blood transfusions. Cancer patients ask that their kidney dialysis machine be turned off. Every competent American has the right to stop any and all medical care that they do not want. *Million Dollar Baby* makes us think that we are more powerless than we really are. Those with severe disability can demand that they be allowed to die; they sometimes do, and they sometimes have their life support stopped.

I don't know if *Million Dollar Baby* will get the accolades it deserves. But I do know that nothing any of those who are criticizing the movie have said should prevent it from doing so. While the movie makes too much of assisted suicide, any movie that can get Americans thinking about life-and-death questions gets my vote.

Physician-Assisted Suicide in Oregon

WHEN IT COMES to Supreme Court appointments, the president and the Republicans in Congress have made it crystal clear what their core requirement is: no "legislating from the bench." With both John Roberts and now Samuel Alito, the president has insisted that he has selected people to serve as judges who will not override the will of the American people. So one has to wonder, What is this administration thinking in pressing the case against physician-assisted suicide in the state of Oregon? Or, more accurately, why is the administration and the president not telling us the truth about how they really see the Supreme Court?

Oregon is the only state in the nation where it is legal for a physician to prescribe a lethal dose of medication to a terminally ill patient who requests assistance in dying. The citizens of Oregon approved "The Oregon Death with Dignity Act" by a ballot initiative in 1994. In 1997, a push was made to revoke the law. Again, Oregonians decided to permit physician-assisted suicide, this time by a larger majority than they had three years earlier.

Various attempts have been made to challenge the constitutionality of the law in court by the Department of Justice. In 2002, U.S. District Judge Robert Jones, in ruling against the attempt by the Bush administration to block the law, said, "Oregon voters decided not once but twice to support the law and have chosen to resolve the moral, legal, and ethical debate on physician-assisted suicide for themselves."

But the administration and the president will not give up. Then–attorney general John Ashcroft pressed the case on appeal, and it has now wound up in front of the Supreme Court. The president's conservative base is so strongly opposed to any form of assisted suicide that it has sought any possible means to overturn the Oregon law. The Justice Department argued before the Roberts court yesterday that the federal

Controlled Substances Act gives the attorney general of the United States the power to prohibit the use of drugs in assisted suicide, regardless of state law. This is truly grasping at a straw to overturn the will of the people of Oregon.

Now, this six-year jihad against the Oregon law might make some sense if there had been a pattern of terrible abuse of the dying and disabled since its enactment. As it happens, I am very wary of legislation permitting physician-assisted suicide. I worry that it could lead to pressure being put on people to end their lives prematurely or people with psychiatric or physical disabilities being dispatched for the convenience of others or to save money. But there is no such record of abuse.

While some Oregonians dying of cancer or AIDS or Parkinsonism do request a lethal dose of medication, very few actually wind up using it. There have been fewer than three hundred cases in the years since the law was implemented. And despite a concerted effort by opponents of physician-assisted suicide to find cases where the law has led to abuse or misuse, I know of only one case in the past five years where any serious challenge has been raised about the ethics of patients, families, or doctors who have honored a request to die.

So what is the president thinking? Why is the Justice Department trying to use a broad interpretation of an obscure federal statute to restrict a law legislated twice by the citizens of Oregon that has not led to any problems or difficulties since its enactment?

There is only one answer. The president is not telling you and me the truth. He is only willing to respect the decisions of Americans if he agrees with them. He is only willing to advocate for a conservative court if it upholds a social agenda that he agrees with. He is not willing to allow a state to follow a policy about dealing with the terminally ill if he does not agree with it. And he expects the Supreme Court to "legislate from the bench" when it suits his moral agenda.

The federal government should not have brought the case against Oregon's law. And the Supreme Court should not listen to the cockamamie argument that a statute intended to prevent the illicit use of drugs gives the federal government the right to tell the citizens of Oregon how they must die when they are terminally ill. The administration constantly bemoans the fact that *Roe v. Wade* imposed a policy on the American people about abortion that was never legislated. Oregon has a policy on assisted suicide that was legislated—twice. The administration and the president should be ashamed for trying to use the Supreme Court to do what they say they do not want any federal judge or court to do. The ethical hypocrisy involved is beyond description.

Lessons from Terri Schiavo

WE HAVE NOW reached the endgame in the case of Terri Schiavo. Her husband Michael remains unwavering in his view that she would not want to live in the state she is in. Despite the fact that he has been made the target of an incredible organized campaign of vilification, slander, and just plain nastiness, he remains unmoved. Even a pathetic effort to bribe him into changing his mind with the offer of a million dollars did not budge him. He say he loves his wife and will do whatever it takes to end an existence that he believes she would not want to endure. He thinks that she would want her feeding tube stopped and that she would wish to die rather than remain bed bound in a nursing home in a permanent vegetative state for the rest of her days.

The Schindler parents and their other children remain equally convinced that Michael is wrong. They say that Terri would want to live, that she is not as brain damaged as Michael contends, that there is still hope for her recovery despite the fact she has failed to show any real improvement in sixteen years, that there remain treatments to be tried, and that as a Catholic, she would want to honor recent papal teachings that feeding tubes should not be removed from those in permanent vegetative states.

Congress, or at least the prolife constituency in the House and the Senate, are doing their best to halt the death of Terri Schiavo. Last-minute bills invoking habeas corpus, a legal doctrine that has historically been used only for those held in federal custody, along with incredibly zany and inappropriate subpoenas to doctors and nurses requiring that Terri Schiavo be brought to Washington, D.C., show a level of grandstanding that is normally reserved for issues such as the use of steroids by major league baseball players.

So now that this miserable case is moving toward a resolution, what can be said about who is right and who is wrong? And what is the likely legacy of the battle over the fate of Terri Schiavo?

Ever since the New Jersey Supreme Court allowed a respirator to be removed from Karen Ann Quinlan and the United States Supreme Court declared that feeding tubes are medical treatments just like respirators, heart–lung machines, dialysis, and antibiotics, it has been crystal clear in American law and medical ethics that those who cannot speak can have their feeding tubes stopped. The authority to make that decision has fallen to those closest to the person who cannot make his or her own views known. First come husbands or wives, then adult children, then parents and other relatives.

That is why Michael Schiavo, despite all the hatred that is now directed against him, has the right to decide his wife's fate. The decision about Terri's fate does not belong to the United States Congress, President Bush, Tom Delay, the governor of Florida, the Florida legislature, clerics in Rome, self-proclaimed disability activists, Randall Terry, conservative commentators, bioethicists, or Terri's parents. The decision is Michael's and Michael's alone.

Remember the recent debate about gay marriage and the sanctity of the bond between husband and wife? Nearly all of those now trying to push their views forward about what should be done with Terri Schiavo told us that marriage is a sacred trust between a man and a woman. Well, if that is what marriage means, then it is very clear who should be making the medical decisions for Terri—her husband.

Isn't it true that tough questions have been raised about whether he has her real best interests at heart? They have. But these charges against Michael Schiavo have been heard in court again and again and again. And no court has found them persuasive.

Has there really been careful review of this case? Is Terri really unable to think or feel or sense? Will she never recover? The flurry of activity in Washington and in Tallahassee might make you think not. But that is not so.

There have been at least eleven applications to the Florida Court of Appeal in this case, resulting in four published decisions, four applications to the Florida Supreme Court with one published decision (*Bush v. Schiavo*), three lawsuits in federal district court, three applications to the United States Supreme Court, and nearly untold motions in the trial court. This has got to be the most extensively litigated right-to-die case in U.S. history. No one looking at what has gone on in the courts in this case could possibly deny that all parties have had ample opportunity for objective and independent review by earnest and prudent judges of the facts and trial court orders.

So it is clear that the time has come to let Terri die. Not because everyone who is brain damaged should be allowed to die. Not because her quality of life is too poor for anyone to think it meaningful to go on. Not even because she costs a lot of money to continue to care for. Simply because her husband who loves her and has stuck by her for more than a decade and a half says she would not want to live the way she is living.

If Terri is allowed to starve to death, what next? Undoubtedly there will be efforts to pass laws to prohibit feeding tubes from being taken away from others like Terri in the future. And there may even be efforts made to push right-to-die cases out of state courts and into federal courts. These are bad ideas.

We have had a consensus in this country that you have a right to refuse any and all medical care that you might not want. Christian Scientists do not have to take medical care, nor do Jehovah's Witnesses need to accept blood transfusions. Some fundamentalist Protestants would rather pray then get chemotherapy. Those who are disabled and cannot communicate have the exact same rights. Their closest family members have the power to speak for them.

The state courts of this country have the power to review termination of treatment cases and have done so with compassion, skill, and wisdom for many years. Those who would change a system that has worked and worked well for the millions of Americans who face the most difficult of medical decisions should think very hard about whether Bill Frist, Tom DeLay, Hillary Clinton, George Bush, John Kerry, or the governor of your state needs to be consulted before you and your doctor can decide that it is time to stop life-prolonging medical care.

Part III

Engineering Ourselves

Is Cosmetic Surgery Always Vain?

WHEN THE SUBJECT turns to cosmetic surgery, at least among North Americans, two things are sure to happen: laughter and embarrassment. Laughter, because many of the icons of surgical alteration—Michael Jackson, Janet Jackson, Pamela Anderson—are so distorted as to make us nervous. Or their efforts to supersize their body parts seem downright amusing. Embarrassment, because so many of us either have thought about undergoing cosmetic surgery or have actually done so.

So is cosmetic surgery simply vain, self-indulgent, and therefore immoral? No.

American attitudes about cosmetic surgery are very much a product of American culture. Compare the behavior of presidential candidate John Kerry's apparent use of botox to get rid of wrinkles on his forehead— he denies it for fear that he will soon be the target of both humor and embarrassment—with the behavior of Silvio Berlusconi, the former prime minister of Italy. Berlusconi has a lot to be embarrassed about these days, but getting a face lift did not faze him in the least. He took a month off to get it done and has been proudly mugging the results ever since.

Even more evidence that our Puritanical past is still alive and well when it comes to surgically tweaking yourself to enhance your looks comes from Brazil. Americans who electively undergo the knife deny it. They would no more fess up to a face lift, nose job, or liposuction than they would the commission of a major felony. Not so in Brazil. Plastic surgery clinics appear like Starbucks on every street corner. If you are wrinkled, unhappy with the size of something, or fear some part of you is sagging, the Brazilian attitude is "Get it fixed." Where we see indulgence and vanity, they see only need and reasonableness.

All of which brings me to a conversation I recently had with parents whose child has Down syndrome. We all know the classic look of a kid with Down's. My friends thought that other kids reacting negatively to that look and wondered if they should send their child to a plastic surgeon to have his appearance surgically altered.

There is some evidence that kids, especially teenagers, react negatively to how a kid with Down syndrome looks. There does not seem to be much evidence about how kids with Down's think about how they look. Surgeons themselves are divided about how well they can alter the appearance of a child with this genetic disease.

But suppose there is a surgeon who could transform the appearance of a child with Down to a more "normal" look. Would it be wrong? Some might say it would be fine since a child with Down's has an abnormal or disfigured face. But is that really the case? Couldn't it be said that a kid with Down's has one of a number of looks that human beings are born with? The decision to try to alter the appearance of a person with Down syndrome would be cosmetic, not reconstructive or therapeutic.

I find myself thinking that altering the appearance of a kid with Down syndrome, if that is what the child and the parents want, would be neither vain nor indulgent. And if that is true, then why not admit that botoxing a wrinkled brow or downsizing a large nose, while cosmetic, may be reasonable choices for some to make as well?

While we love to gossip and even snigger at those who use cosmetic surgery to look better, maybe we should lighten up a bit. Although cosmetic surgery, like anything else, can be abused or misused, is it always just unseemly vanity to want to look better?

Face-off over Gene Foods

THE RESIDENTS of the state of Oregon voted way back in November 2002 on a referendum that has important national implications. Proposition 27 would have required that any food or dairy product that is produced through genetic engineering be labeled. A powerful alliance of food manufacturers and agricultural biotechnology companies poured money into the state to ensure that the labeling requirement was defeated. And it was. But the issue of labeling for genetically engineered food is not going to go away anytime soon. This is because, no matter how much money industry has at its disposal to kill this kind of bill, it is hard to argue against the public's right to know about what they are eating.

The critics of Oregon's Proposition 27 did have some very solid arguments. They argued that labeling was not necessary since genetically modified organism (GMO) food is safe. They also maintained that labeling would be very expensive, raising the price of products that were made in Oregon. An argument similar to this has been raised since the Oregon vote by food manufacturers that are faced with labeling requirements in Europe and Asia.

There really is not much room for argument about the safety of GMO foods. The Food and Drug Administration (FDA) weighed in against the proposed Oregon law on the grounds that there is no scientific evidence that foods containing biotech ingredients cause health problems. Genetically modified foods have been used in the United States and other nations for many, many years. It is very likely that you have eaten food made with genetically modified soybeans or corn. Right now 70 percent of processed foods in America's grocery stores contain GMO ingredients. No health problems of any sort have been found that lead to food with genetically modified ingredients. Safety is not an issue. Safety is not a reason to insist on labeling food containing GMO ingredients.

The argument about cost is not as persuasive. Japan, Australia, New Zealand, and Taiwan all have mandatory labeling laws, and the price of food in these countries has not been affected in any significant way. If it is cost that the industry is worried about, then there is hard evidence that shows that it really does not cost much to label.

So, even if it is cheap, why label if there is nothing to fear from GMO foods? The answer is that genetic technology has not yet been used either to improve your food or to reduce the burden current farming techniques create for the environment. To date, little effort has gone into making food healthier, safer, or more environmentally friendly to grow. Instead, companies like Monsanto have used the technology to create crop plants with more resistance to the herbicides and pesticides they already sell. This does little to cement consumer trust. And it gives the consumer no reason to buy GMO-containing foods.

The way to change this state of affairs is to make sure that consumers get what they want from genetic engineering—safer, healthier food that takes a lesser toll on the environment to create than the current methods using pesticides and herbicides. The way to do that is for industry to be pro–consumer information. Then consumers can look for GMO food that is healthier and better for the environment. That is not the position the biotech industry took in defeating the Oregon bill, but it is one that it may well come to regret.

The biotech industry should promise to create an aggressive, open, and thorough voluntary program of consumer information about genetically engineered foods. Anyone who wants to know what foods have genetically engineered ingredients should be able to look them up on a website or by dialing a toll-free number. Anyone who wants safety information on foods that have been genetically engineered (and remember the safety news is good) should know exactly how to get it. Anyone who wants to know what they are eating in a restaurant or cafeteria should know. And anyone who wants to ask why, to date, genetic engineering has been used only to make food pesticide friendly rather than environmentally friendly should be able to ask. The only way to improve what you eat is to make sure that those who control the technology keep no secrets about what it is that you eat.

Heightened Questions about Growth Hormone

THE FOOD AND DRUG Administration not so long ago made a decision of enormous significance. The agency decided to allow the use of Humatrope, the Eli Lilly company's growth hormone, for use in small children. The reason this was big news is that this is one of the first times the agency has allowed a drug to go on the market to treat something that is completely normal. It is unlikely to be the last.

Humatrope is given by injection five or six times a week. It is not cheap—$20,000 to $25,000 a year on average. There is a small risk of side effects—headaches, muscle pain, and even some rare forms of cancer have been linked to the drug. Even more troubling, not every kid who gets the drug gets bigger. On average, after five years of shots, kids who respond show between one and a half and two inches of growth.

Being short is in itself not a disease. But parents, especially of boys, get frantic when they think their kids might be at the very low end of the normal growth curve. So, despite its cost and the fact that it is hard to see what difference an inch and a half will really make, there is likely to prove to be a very big market for Humatrope.

One way to cope with being very short is to take a drug to grow taller. Another is for parents to build confidence in their kids despite their height. Still another remedy for being very short is to teach other kids, short and tall, that height is just a difference among us, nothing more.

But parents don't have to take the shots. And they worry what will become of their kids in a world where height is most certainly rewarded. And so, despite the promises from Lilly that it won't hype Humatrope, there is likely to be a rush to use it in boys who are at the low end of the growth curve. Those who are small are likely to show the way toward a future in which medicine plays a bigger and bigger role in trying to eliminate any condition that is unusual or stigmatized. Bigger may not be better, but the pursuit of better is likely to get a whole lot bigger in the years to come.

Brain Enhancement

THE REVOLUTION in our ability to identify and manipulate genes has spurred all sorts of ethical debates, but an equally profound revolution—in brain science—is attracting amazingly little attention.

Progress in many areas of neuroscience promises not only to reveal how the brain works in general but to provide information about our intentions, thoughts, and feelings as well as the mental aberrations that plague so many of us. Right now sophisticated imaging tools are enabling scientists to see which parts of the brain are active at any given time and to observe the effects of drugs, fear, or other stimuli. Granted, researchers so far have only a limited understanding of brain function. But that will change. Because the structure and activities of our brain influence our mental health and behavior much more directly than our genes do, it is very likely that advances in the ability to "read" the brain will be exploited as much as, or more than, knowledge of genetics for such purposes as screening job applicants, diagnosing and treating disease, determining who qualifies for disability benefits, and, ultimately, enhancing the brain.

Already lawyers are attempting to submit brain scans as evidence of their clients' innocence. Government agencies are considering scanning the heads of prospective military pilots, astronauts, and secret agents to see who might be predisposed to do what in response to stress or temptation. Doctors are implanting devices directly into the brain to help patients cope with Parkinson's disease. There is talk of pills to aid soldiers in erasing the memories of war horrors and implants that might repair or even enhance memory. And high school kids who have no obvious learning disabilities are swallowing Ritalin and other psychoactive drugs to get an edge when they take classroom exams or SATs. All this activity ought to get our ethical radar going, prompting considerations of who might be harmed and how to protect those people. At the moment, questions far outnumber answers, but identifying key ethical issues is an essential first step.

It is hard to argue that anything is fundamentally wrong with trying to detect and ameliorate brain disease. But those efforts nonetheless raise serious issues similar to those arising from the ability to perform genetic testing and therapy: Who decides, and on what basis, whether a risky procedure is justified for a given person? And does each of us have the right to insist that no testing or intervention be done, with no results shared with others, unless we give our consent?

Even people comfortable with the idea of fixing obvious brain deficits become much prissier when it comes to mucking with brains to make them better than good. Americans in particular believe that people should earn what they have. Having a brain that can do more as a result of a drug or a chip or an implant seems like getting something for nothing.

But is it really so terrible if techniques used to treat Alzheimer's disease or attention-deficit disorder lead to ways to improve normal memory? Would it be bad if some innovation—say, a brain chip implanted in the hippocampus—enabled a person to learn French in minutes or to read novels at a faster pace? Should we shun an implant that enhances brain development in newborns? If altering the brain makes it possible to perform better, achieve more, or have greater capacities than one's parents, is such alteration patently immoral?

Unfair? Unnatural?

I see little wrong with trying to enhance and optimize our brains. To clarify why I think this way, let us consider some likely objections.

One is that tweaking neurons to improve brain capabilities might threaten the equality of human beings; by becoming advantaged, recipients would achieve more and command greater respect. But the right to be treated with respect has never depended on biological sameness or on a leveling of behaviors. Just as the disabled and sick should never have a lesser right to fair treatment, happiness, and opportunity, neither should those who do not receive brain-enhancing interventions.

Many people believe that enhancement would be unethical because some of us would be able to get an improved brain and some would not, which would be unfair. It is certainly possible—in fact, probable—that if nothing were done to ensure access to brain-enhancing technologies, inequities would arise. But as Kaplan test preparatory courses, music camps, and math tutors remind us, access to things that improve the mind is already skewed unfairly. This state of affairs does not make inequity right. The solution, though, is to provide fair access—be it to teachers or implantable chips—not to do away with the idea of improvement. As it happens, my son is privileged; he goes to private school. If I told people in a poor neighborhood about this education, they would not say I should be ashamed of myself for giving him an advantage. Nor would they claim that better education is immoral. They would say, "I wish I could do that for my child."

Equity aside, isn't it true that brain engineering is unnatural? If we started to enhance ourselves, we might be able to do more, but would we still be human when we were done? The main flaw with this argument is that it is made by folks who wear eyeglasses, use insulin, have artificial hips or heart valves, benefit from transplants, ride on planes, dye their hair, talk on phones, sit under electric lights, and swallow vitamins. What are they really talking about? Have we become less human because we ride instead of walk to work? We might be less healthy, but does a reliance on technology for transportation make us unrecognizable as humans? Is there a natural limit beyond which our nature is clearly defiled by change? Surely not. It is the essence of humanness to try to improve the world and oneself.

Last, some may argue that brain enhancement is wrong because it will inevitably involve coercion. Subtly or otherwise, the government or corporate advertisers will convince us that unless we have the best brains possible, we will be letting down our families and communities. People might also feel coerced in that if they did not submit to enhancement, they might be left behind in the hunt for jobs and social success. But the answer is not prohibiting improvement. It is ensuring that enhancement is always done by choice, not dictated by others.

In reality, though, it is unlikely that coercion will be needed to induce people to want to optimize their brains. Market-driven societies encourage improvement. Religious and secular cultures alike reward those who seek betterment; every religion on the planet sees the improvement of oneself and one's children as a moral obligation. If anything, the impending revolution in our knowledge of the brain will require us to build the legal and social institutions that allow fair access to all who choose to do what most will feel is the right thing to do.

Seasonale: Medicine for the Sake of Convenience?

IS THERE ANYTHING about the human body that medicine should not try to alter? When it comes to women's bodies, the answer apparently is no. Medicine is more than ready to fool with Mother Nature.

A study reported that there has been a 20 percent jump in elective Cesarean sections. These are not C-sections chosen by women when natural childbirth proves too difficult. Rather, these are surgeries for women who are scheduling them simply for the sake of convenience.

As one mother-to-be told one of the researchers, "Our vacation is important to the entire family, and I would rather have the birth over with than ruin that for everybody."

Elective C-sections are not the only recent example of medicine being used to undermine the natural. The Food and Drug Administration decided to allow a new pill to be put on the market that can reduce the number of menstrual cycles women experience. Instead of having their periods once a month, women who take Seasonale would have only four periods a year—one in the spring, summer, fall, and winter. Get it? Seasonale!

Doctors have long known that women who take birth control hormones do not menstruate. The reason that birth control pills are supposed to be taken for twenty-one days with a break for a period is that the original manufacturers wanted to make birth control pills seem more "natural." They also knew that having a period allows a woman who is trying to avoid pregnancy know that she is not pregnant.

But today birth control pills already seem "natural." And there are easy-to-use test kits to see if you are pregnant. So who needs menstruation?

While many women would like to be rid of the inconvenience of periods, should it be part of medicine's job to help women time their births to fit a busy schedule or to get rid of a messy and sometimes painful monthly experience?

Like it or not, the answer seems clear: Say good-bye to menstruation.

Most doctors would agree that C-sections are worse for moms and babies than natural childbirth. But that concern has not slowed the explosion of elective C-sections. And, despite the fact that there are no long-term studies on the side effects of suppressing menstruation, Seasonale is now on the market.

Just as elective C-sections become the norm, the advent of Seasonale illustrates how medicine is increasingly willing to let the patient decide what is and is not appropriate when it comes to taking health risks.

And it's not just about patient's choice. As long as there is big money to be made doing elective surgeries, the rate of elective C-sections is likely to continue to climb. Similarly, you can bet there will be a big market for Seasonale. There is a lot of money to be made selling it—and few women are going to wax nostalgic about missing their period.

Medicine is poised to change the very nature of what is natural. When money and convenience are allied with what medicine can offer, Mother Nature does not stand a chance.

Raffy and the Trouble with Steroids

PEOPLE WHO DON'T tell the truth are very good at lying to your face. Anyone who has watched the truth shaving of such well-known confabulators as Ken Lay, Bill Clinton, Bernard Ebbers, Scooter Libby, and Charles Keating knows what I mean. Rafael Palmeiro has joined the ranks of those who will be remembered as being good at delivering the big lie.

Raffy, as those in the know in the world of sports refer to him, has proven positive for steroid use after sitting in front of a congressional committee denying unequivocally that he was a "juicer." This rapid fall from grace of one of baseball's heroes, a three thousand hit and five hundred home run candidate for the Hall of Fame, has left sportswriters fuming. Raffy seemed to be a guy you could trust. But, as anyone who has to deal with addicts knows, truth evaporates when it comes to their drug of choice.

Now the airwaves and blogs burst with debates about whether Palmeiro has succeeded in lying his way right out of Cooperstown. There is, however, a much bigger and more fundamental ethical question raised by the Palmeiro implosion and the steroid scandal that continues to haunt major league baseball.

It is easy to condemn steroid use. The drugs, while effective, are very dangerous. But what if they were not? How are professional and amateur sports going to deal with the impending explosion in performance-enhancing drugs and bioengineering tricks that can boost performance with little or no risk for the user?

At my school, the University of Pennsylvania, physiologist Lee Sweeney is hard at work trying to find ways to tweak genes to make muscles grow bigger and more densely. This research holds out real hope for those with muscular dystrophy and other debilitating muscle diseases. But the gene transfer technology he is working on will also make it possible to make normal muscles bigger and stronger, too. Figuring out who may or may not have engaged in "gene doping" will prove next to impossible. And it is likely that there will be little risk associated with genetically altering muscle cells.

Similarly, scientists around the world are busy making pills that will let us sleep better, control tremors, fight fatigue, slow the loss of memory, speed up learning, recover from hard exertion more quickly, and calm anxieties. Some of us are already benefiting from drugs like these when we use Ambien, Provigil, Ritalin, Prozac, or Effexor.

So what are we going to say when the archer, the chess master, the competitive marksman, the Nascar driver, or the woman's professional golfer says, "If I take these same drugs, I just might get enough of an edge to move ahead of my competition"?

Throughout the nineties—when home runs were flying out of baseball stadiums launched by players who were obviously using steroids, when professional football linemen got huge, when track and field records continued to fall—not much in the way of protest was heard. Americans are in love with those who take risks to break a record or each other's bones in the name of sport.

Nor do Americans gripe when we show up at the Olympics with our athletes who have the best training, superb diets, and top-flight equipment and whomp the tar out of athletes from poor nations, some of whom seem to have shown up just to get a decent meal. We are used to employing science to our advantage when it comes to sports, so why should we draw the line at genetic engineering or new miracle pills?

There is nothing about the reaction to Rafael Palmeiro's downfall that indicates we are ready to deal with the fundamental ethical question raised by his use of steroids: is there any place in sport for enhancement? Is the point of sport to see what human beings can do without aid of any sort in fair competition? If so, we may need to close the training facilities and cut back on what dieticians and trainers are allowed to do. If the point of sports is to test the limits of human performance, then we had better get ready to add genetic engineers and a bevy of pharmacologists to the hordes of specialists now working with elite athletes from elementary school to the pros. There is no right answer to what the point of sport is. But Rafael Palmeiro has made it a question no one who cares about sports can avoid.

When Steroids
and Politics Mix

IT'S ALMOST TIME for Californians to decide whether to keep Arnold Schwarzenegger as their governor. But Schwarzenegger has an ethical problem that makes him unfit to serve, an issue that has stared his supporters in the face ever since he first took off his shirt.

No, it is not the fact that Schwarzenegger made his fortune selling violence on the silver screen that makes him unfit for office. Or his numerous, well-documented sexual escapades.

The baggage weighing down the Hollywood he-man is that for many years he used steroids—drugs that are illegal in the United States today. And those who normally yell the loudest about drug use in America have kept mum because the former Conan the Barbarian represents their best shot to hold political power in the nation's most populous state.

For most of his life, Schwarzenegger has been either a bodybuilder or an actor. In both occupations, his sole talent has arguably been his big muscles. And the only way he came to have the massive form that made him famous was through the use of steroids.

There is probably not a doctor or athletic trainer in the country who believes you can look like Schwarzenegger without a chemical boost. His earlier use of these drugs inspired—and continues to inspire—lots of young men to risk their lives by emulating what he did.

Does anybody really think that the mass and size Schwarzenegger was famous for in his movies were simply the products of a good diet and hard work in the gym? Not a chance.

Yet, conservatives, who are quick to scream for a harsh jail sentence whenever an NBA airhead is arrested for drug possession, are touting Schwarzenegger as the best man to lead the nation's biggest state.

Furthermore, neoconservatives, such as William Kristol, Bill Bennett, and Charles Krauthammer, who have been frothing at the mouth in op-eds and magazine articles about the immorality of using new biological techniques to redesign our bodies, have had nothing to say about the once pharmacologically enhanced Schwarzenegger.

Keep in mind that the health consequences of steroid use are not trivial. Liver tumors, cancer, hypertension, torn tendons, and sterility are strongly associated with the use of these drugs.

As the U.S. Customs Service says on its website, "Steroids are like any other illegal drug that threatens the American public—like all illegal narcotics their sale and possession represent critical links in a larger criminal process, one that funds terrorism, death and addiction around the world."

And, as the National Institute on Drug Abuse notes on its website, some of the greatest abusers of steroids are junior high and high school students who are using these drugs at rates 50 percent higher than they were ten years ago. And who might these kids, now desperate to bulk up their bodies, have been watching on TV and in the movies for the past twenty-five years?

A lot of the folks supporting Schwarzenegger for governor had no doubt that the White House was no place for a philanderer who lied about his sexual behavior. I agree. But these same people ought to check their moral compasses and conclude that the governor's office in Sacramento is no place for an unrepentant drug abuser.

ANDi the Fluorescent Monkey

THE SCIENTISTS who made ANDi at the Oregon Regional Primate Center had to be a bit disappointed that he does not glow in the dark. He could have been a major media star. Proof that ANDi really is the first primate to be genetically engineered can be found only by using a microscope to see his cells glow.

So what are we to make of this? Does it really matter that scientists can make a slightly fluorescent monkey? How much of a demand is there, really, for glow-in-the-dark cats, dogs, or wayward kids out too late at night on their bikes, anyway?

Probably not much. But ANDi represents something much more important. The tiny light cast by this baby monkey shows that it is possible to genetically engineer ourselves.

The scientists in Oregon took a tiny step toward doing what many scientists have said no scientist would ever want to do: use genetics to change, improve, or enhance our children. Sticking genes into eggs and growing a healthy monkey means that someday scientists could and most likely will insert genes into human eggs that make kids smarter, stronger, faster, healthier, or happier than their parents.

So, is the prospect of a Catherine Zeta-Jones with the mind of a Stephen Hawking something we should celebrate or outlaw?

Some will surely argue that we need tough laws to prevent some kook from setting up a DNA shop on a deserted island and breeding superbabies—a genetic Temptation Island. Others will say that we need an international ban lest we find ourselves taking orders from the next Saddam Hussein's eugenically brewed army.

No such laws are needed. Renegade scientists and totalitarian loonies are not the folks most likely to abuse genetic engineering. You and I are—not because we are bad but because we want to do good.

In a world dominated by competition, parents understandably want to give their kids every advantage. And there is hardly a religion on the planet that does not exhort its believers to enhance the welfare of their children. The most likely way for eugenics to enter into our lives is through the front door as nervous parents, awash in advertising, marketing, and hype, struggle to make sure that their little bundle of joy is not left behind in the genetic race.

Most parents are willing to spend a lot of their money sending their kids to colleges, getting them piano or tennis or language lessons, making sure that they eat well and are safe. There is little reason to think that the drive to do right by our kids will be any different if and when we are offered the chance to improve them genetically. No one will have to fool us or force us—we will fall over one another, eager to be first to give Junior a better set of genes.

The antidote to the blind application of genetic engineering is to start talking about what should and should not be allowed, who will pay, and what standards ought to apply to those who want to advertise and sell utopian children. The right response to ANDi is not legislation to stop nuts but conversation that will teach us how best to control ourselves.

Part IV

Engineering Plants, Microbes, and Animals

Are Genetically Modified Foods Fit for a Dog?

THE FRENCH agricultural cooperative Cana-Caval eliminated genetically modified ingredients from a large portion of the pet food it makes. This follows hard on the heels of a similar decision by the giant French supermarket chain Carrefour to ban genetically modified ingredients from its store-brand pet foods. The soy, maize, and other grains added to Fifi's dinner will be subjected to high-tech genetic testing to weed out any genetically offensive material.

It is true that the French have always followed their own compass when it comes to many things, especially food. But spending money to genetically screen whatever it is that makes it into the dishes of dogs and cats in France nicely illustrates how the debate over genetically modified (GM) food has lost its direction.

The main worries about GM food seem to be either that it is unsafe to eat things that have been genetically engineered or that growing genetically engineered food poses dangers to other animals and plants. The latter concern does merit serious consideration; the former, with all respect to French dog food manufacturers and their customers, much less so.

Past experience with new animals and plants, ranging from the African honey bee to the Japanese beetle to any number of other pests roaming around the world, shows that the introduction of new creatures into old environments can lead to big trouble. Growing any new plant, genetically engineered or not, requires great caution lest something new winds up unintentionally replacing something old.

There is no excuse for not having better rules and governance for the introduction of new genetically engineered species than currently exist. Today, there are too many voluntary guidelines and even less rigorous enforcement. The introduction of new species in the past proves that good intentions can lead to bad outcomes, so more must be done to control the introduction of new forms of GM seeds and plants.

Eating is another matter. GM food has been around in the food supply for many years. Americans and those in many other nations as well as their pets have been snarfing down huge quantities of GM soy and other grains in many, many foods. No one has yet seen any cats with four tails or people with two heads.

Those who would not feed genetically modified products to a dog maintain that the fact that no bodies are piled up outside the GM food plant proves nothing. GM food may still kill you in the decades to come.

Perhaps, but in a world where so many things known to be bad for you are consumed with abandon, it is hard to worry about the risks of foods that have an exceedingly remote possibility of being harmful. A large part of the developed world eats so many bad things that they and their pets die prematurely due to obesity. A huge percentage of the world dies simply because they do not have enough to eat.

The wonderful promise of genetically modified food is that it can fix both problems. Genetic engineering can get more soy into our diets while getting rid of the bad fats and can also increase the amount of food available so that the poor need not starve.

The French may not think GM food fit even for animals. They are wrong.

Grown safely, GM food will be the solution to the current food crisis in which too many die from having too much that is known to be bad, while too many die from having no food at all.

Miss Cleo, Meet "CC,"
the Kitty Clone

WHAT DO GENETICS Savings and Clone, the company that recently announced the cloning of CC, the calico kitten, and Miss Cleo, the late-night TV psychic with the over-the-top Jamaican accent, have in common? Both are in the business of preying on human weakness.

Miss Cleo is currently in a whole lotta trouble. The state of Missouri is throwing around legal threats after receiving many complaints from irritated denizens that when they call Cleo to find out who is trying to seduce their husband, not only did they not get to speak to her, but they also wind up with huge telephone bills.

Meanwhile, the Federal Trade Commission has been emitting unhappy noises about the overnight soothsayer, stating that the operation run by Miss Cleo and her employers, Access Resource Services Inc. and Psychic Readers Network, is permeated with fraud.

Florida authorities have filed a separate lawsuit challenging Miss Cleo, whose real name is Youree Dell Harris, to prove that she really is a renowned shaman from Jamaica.

As Miss Cleo might say, times is hard at the hotline.

Times do not appear to be quite as hard at Genetics Savings and Clone. The company says that folks are lining up to hand over $50,000 to $100,000 to clone their deceased pets.

But Genetics Savings and Clone is selling a product—immortality—that it can no more deliver on than can Miss Cleo peer into your future.

When the creation of CC, the first cloned cat, was announced, something puzzling was immediately obvious. The cloned cat looked different from its parent. The folks at Genetics Savings and Clone could not have been pleased. CC is a living reminder that cloning is not copying. You don't get an exact replica of the animal. You get the same genes, but any animal, including you and me, is more than its genes.

Environment, including that in the womb, plays a big role in determining the color pattern of calico cats, and the environment is not duplicated by cloning. The personality, behavior, and traits of any cloned animal will depend on more than genes. The new cloned cat will not know the old cat's tricks.

Pet owners may dream of the day when their beloved pet can be restored to them good as new. But it is only a dream. Cloning cannot cheat death. It can only provide you with a new pet that is likely to look enough like the old one to remind you every day of the differences between the one you have and the one you had.

Genetics Savings and Clone does not cotton to the idea that those who are willing to spend considerable sums of money to get their dead pets back are being scammed. The company says it tries hard to make sure its customers know that cloning is not resurrection. But these protests about informed consent and educating the consumer ring no more true than Miss Cleo's hotline.

The only way you can get people to pay you a lot of money for telling fortunes or cloning their pets is to have them believe in something that is not true. It isn't right when Miss Cleo and her corporate sponsors prey on the hopes and dreams of the gullible to make a buck. It is no better when corporate science preys on the hopes and dreams of the grieving.

Whipping Up the Avian Flu

AVIAN FLU IS much in the news these days. And it should be. This is one nasty little critter. Why, then, would a recipe for how to make a nasty version of it appear in a leading scientific journal where anyone, including some of our worst enemies, can find it?

How bad is the current version of avian flu now racing toward us from Asia and Europe? Compare the impact of the avian flu on your lungs with the regular old strain of flu that appears around this time of year—it takes your breath away.

Literally.

Avian flu releases fifty times as many infectious particles in the human lung as does an ordinary flu virus. If you wait four days, there are thirty-five thousand times as many virus bits in a mouse lung infected with avian flu than are present in good old normal but nasty flu. In mice, 100 percent are dead a week after infection, compared with a few deaths from other flu viruses.

When influenza first struck globally in 1918, it killed as many as fifty million people worldwide. Pandemic avian flu will probably kill just as many people, if not more.

Even worse, the existing flu vaccine does not protect against avian flu. Prescription medicines like Tamiflu may not do much if you do get infected, and a naturally occurring strain of avian flu resistant to Tamiflu has already been discovered. About the best you can hope for is not to get infected, which may not be an easy thing to do at a time when modern air travel means someone can be infected in Romania, Thailand, or Indonesia on a Monday and be standing next you on a street corner by Tuesday.

Given this grim picture, you might imagine that the last thing scientists would try to do is to create this or a similarly lethal bug in the lab. You would be wrong. A team of scientists recently announced in *Science* magazine they had re-created an artificial version of the original pandemic flu virus.

53

At a time when the president and the administration are making the war on avian flu a top priority, why are scientists making a similar deadly virus in their labs?

One explanation is that, by building this bug, it may be possible to understand every bit of its genetic blueprint. That information could prove invaluable both in identifying any naturally occurring viruses and in developing a vaccine against avian flu.

However, it is not at all clear that the re-created virus is being kept in such secure confinement that a bioterrorist could not get his or her hands on it. Worse still, publishing the entire recipe for how to make a deadly virus gives every terrorist group and nutball outfit in the world the opportunity to try their hand at making the virus, too.

The state of the art in science is that scientists have only truly accomplished something when they prove it with publication in carefully peer-reviewed publications. Putting the virus out there for peers to judge was the logical next step along the path to understanding and controlling the machinery that fuels a potential pandemic.

Those in the life sciences hate the idea of government meddling in their work. They believe the free exchange of information is the best defense against a nasty pathogen like avian flu, whether it gets here on an airplane or in a terrorist's bomb. But can we really be sure that every step has been taken to keep existing samples of flu secure? And does it really make any sense to publish the complete genetic blueprint for the virus where anyone can find it? Perhaps not.

A decade ago, the manipulation of deadly viruses could be restricted to a high-security vault, a hot zone under lock and key. No one had anything to fear from a scientist's description of a virus. Today, however, the supersecret stores of deadly viruses have been reduced to a set of instructions, which might at some point become a cookbook for terrorists and other malcontents or amateurs.

At a time when there are no means to easily prevent the spread of avian flu and no therapy to treat anyone who gets infected with it, there is a need for much more accountability about who should be making this and similar organisms. And we need to be especially vigilant about who they should be telling about how to make it.

Should Scientists Create New Life?

SHOULD SCIENTISTS try to create a new life form? That profound question was put on the nation's agenda when two of the world's premier geneticists, Craig Venter and Hamilton Smith, announced that they had received millions of dollars in federal funding to attempt such a project.

In a nutshell, what Venter and Smith propose to do is create a very simple microbe. Some viruses have only four hundred or five hundred genes. Venter and Smith are going to remove the genes from one such tiny bug, synthesize a new set of genes, drop them into the bug, and see if the new instructions will bring the microbe to life. If they do and if the genes are a blueprint that has never existed before in nature, then these two scientists can say that they have created the world's first artificial microbe.

Aside from the glory involved, there are some very good reasons to build artificial viruses and bacteria. It would be of enormous importance to know what sequences of DNA put in the right order can make something come "alive." Such knowledge could one day enable us to make microbes that we could use to prevent pollution; kill other bugs that cause us much misery, such as those involved in gonorrhea, malaria, tuberculosis, and AIDS; and make tiny microbes that could provide some immunity against disease.

But there are some reasons to think hard about the wisdom of trying to create new life. Some will worry that it is not our place in the cosmos to create living things. Only God should do that. Others may fear that the creation of new microbes holds the potential for having something escape from the lab that could cause havoc out in the real world. Still others correctly note that this kind of synthetic genomics could be used to make nasty critters that terrorists or evil nations could use against us.

It might be noted, too, that those who can build a bug can also claim ownership over it, making it possible that someone might actually try to patent "life" itself!

I don't think we should fear the creation of new life forms. After all, we have essentially been doing that for many centuries through systematic breeding of animals and plants.

As far as I can determine, no major religion is opposed to the creation of life forms as an act that defies God's will or places humanity in a role that is inappropriate. Playing God is a common criticism of what scientists do, but if you create a life form that can cure disease or feed the world, then you won't find many religious leaders objecting.

And yes, nasty microbes may escape, and bad guys could synthesize some pretty nasty critters, but what we need here are adequate safeguards and controls. Perhaps not every bit of information about how to make a microbe belongs on the Internet or in a publicly available journal. And if we don't want anyone to own new life forms, then it is well within our power to pass legislation that would prohibit them from doing so.

So, amazing as the idea is of generating new life from scratch, I think there is merit in pushing ahead. But if this ship is going to sail, we have to get busy making the rules and the safeguards as well as the genes.

Smart Mice, Not-So-Smart People

THERE WAS GOOD news for mice and bad news for humans recently. A team of biologists announced they had used genetic engineering to create a smarter strain of mouse, a feat that opens the door to boosting human intelligence. But in Kansas, school officials voted to remove evolution from the state's required science curriculum—not a smart move.

Joe Z. Tsien at Princeton University and his team created genetically enhanced mice that showed themselves to be smarter than normal mice by performing better on a set of standard tasks, such as recognizing new objects or finding ways to get out of a bucket of water.

Unfortunately, the kind of genetic engineering that proved successful in mice is not ready for deployment in humans. Tweaking human genes involving memory would not necessarily lead to the kind of increase in intellectual performance that Tsien's mice exhibited. Most of us need a more varied form of smarts than knowing how to escape from a bucket. Moreover, risks to the health of mice brought on through unanticipated changes that this new kind of genetic engineering may cause would not be acceptable in a human baby.

I say it is unfortunate that we are only on the threshold of knowing how to enhance human intelligence because some sort of enhancement is in order for some of those living in the state of Kansas.

While biologists were using their knowledge of evolution and genetics to engineer a smarter mouse, the state school authorities in Kansas decided that kids there no longer need to be tested about their knowledge of evolution. This is really stupid.

Ever since Charles Darwin put forward his theory that the world around us is the result of slow change caused by natural selection acting over many eons, a few of those who hold to literal accounts of creation as presented in the Bible have been on the attack. Some have taken the implausible view that creationism is somehow an alternative scientific account to the theory of evolution that has evolved over the past century and a half since Darwin.

Whatever creationism is, it is not science. Indeed, it is an insult to religious belief to hold that the creation account presented in the Bible is somehow susceptible to the same sorts of tests and challenges that fallible human reasoning is.

The work that led directly to the creation of a smarter mouse is based solidly in evolutionary theory. It is not possible to map the genome, clone animals, breed new species of grapes or soybeans, or undertake gene therapy on a baby without relying on the validity of evolutionary theory. That is not to say that evolution is the last word about how we and mice and the rest of nature all got here. It is entirely possible to fervently believe in the literal truth of the Bible and to nonetheless understand that human beings have created a view of the world that can be believed whenever one is thinking like a scientist.

The residents of Kansas need to get smart. If mice can be made smarter by rearranging their genes, then kids in our high schools and colleges need to know how and why scientists are able to do this.

If we are going to cope with the genetic revolution that looms before us, then every American needs to be taught about evolution and genetics— but not because it is the only way to view the mysteries of the world around us.

If we fail, mice smart enough to know the science behind their creation may well displace us in the classroom.

Part V

Experimentation Ethics

Testing Biological and Chemical Weapons: Any Volunteers?

THE THREAT THAT biological and chemical weapons might be used in the context of armed conflict or by terrorists is frighteningly real. Advances in genetics, microbiology, bacteriology, and cognate areas of biomedicine are presenting a new array of threats to both military personnel and civilians. Synthetic virulent strains of viruses and bacteria, "fusion" toxins, and stealth viruses—along with novel modalities for delivering toxic agents using organic and inorganic chemicals, foods, aerosols, and microdroplets—have raised the concern of federal and state governments, national security agencies, and the armed services. Research efforts, supported by significant federal and private funding, are under way to find vaccines, drugs, prophylactic agents, and palliating interventions that might mitigate these threats.

The explosion of research in this area has created an important but little-discussed ethical challenge. Many institutions are struggling with issues raised by the design and oversight of research protocols that call for the deliberate exposure of human subjects to toxic, noxious agents. Some of this research will of necessity be done secretly despite the fact that, as the Final Report of the Advisory Committee on Human Radiation Experiments noted nearly a decade ago, this poses a tension "between duties to disclose and the need to keep information secret" (available at http://tis.eh.doe.gov/ohre/roadmap/achre/report.html). Federal regulators and scientific and medical journals will soon be asked how to ensure that such experiments are carried out in a way consistent with the highest standards of ethical conduct in the protection of human subjects.

Contemporary human subjects protections apply almost exclusively to research that seeks to produce generalizable knowledge that can be put to beneficial use in biomedicine—for example, by creating new diagnostic tests, new therapies, or new forms of prophylaxis against naturally occurring diseases. Review bodies and regulators assess informed

61

consent and risk–benefit ratios in the context of research to improve health. There is little experience with experiments that deliberately harm subjects.

Such experiments have, for the past two decades, been limited to three areas.

The first is research on treatments for common, nonlethal viruses, which requires first infecting subjects. The risks to subjects in these types of studies are known to be very, very small. The second is "phase 1" clinical trials, done for assessing the safety of new drugs. The third is challenge studies in which basic physiological and psychological information is sought about various stimuli. Current ethical standards require that subjects in such studies be closely monitored and that risks be kept to a minimum.

Research involving biological or chemical weapons must of necessity involve exposures to toxic agents and levels of risk higher than those that exist in most research. Such research would seem to compromise the core ethical tenet of medical ethics that studies should not knowingly do harm. Indeed, the Environmental Protection Agency now refuses to accept most toxicity tests done on human subjects in order to establish "safe levels." Yet, national security now places a higher premium on studies that might pose similar risks, and many prospective subjects might wish to volunteer from a sense of duty, patriotism, or for personal financial gain.

There is every reason to believe that this work will move rapidly forward. If so, clear guidelines are needed for establishing its unequivocal relevance to national security concerns. These guidelines must also address who may be recruited as subjects, what level of competency they should demonstrate, how the freedom of their choice can be assured, what types of end points will be used, what compensation they will be given, and what level of oversight will be in place. Investigators, review committees, and journal editors will need guidance about the kinds of harm that can be associated with this research and the ways in which restrictions on the dissemination of research results should shape their assessment of its morality.

Some will surely argue that no form of research involving the deliberate harm of human subjects ought be tolerated. Such a policy could be put in place. It would have obvious implications for the speed with which antidotes to biological and chemical weapons can be found and for the confidence those receiving them can have in their efficacy. If, on the other hand, the need to hasten discovery in this area leads us as a nation to permit such research, then it is imperative that the norms needed to ensure that it is conducted fairly and humanely be formulated, then widely discussed, and then thoroughly debated as soon as possible.

Lawsuits Are Not
the Answer

IF YOU LIKE the impact that malpractice has had on American medicine, then you are going to love what it is about to do to human experimentation. If nothing is done, a lot of doctors and researchers are likely to decide that there are less costly ways to make a living than doing research on diseases and disabilities.

A lawsuit in a human experimentation case that generated a great deal of local and national attention wound up being settled. James Quinn died at the age of fifty-one at Hahnemann University hospital in Philadelphia. He was the fifth person to receive an experimental artificial heart. The machine was put into his chest on November 5, 2001, when he had only a few more days to live. Quinn did better than expected. The machine kept him alive for nearly a year, until August 25, 2002, when he suffered a stroke and the decision was made by his surgeon to shut off the device.

Quinn's widow was angry about how her husband had died. She felt that he never would have consented to have the device put in if he had known that it would make the last few months of his life miserable. She also maintains that neither she nor her husband knew just how experimental the artificial heart was. They expected the machine to save him, not just to give him more life followed by a miserable death.

Mrs. Quinn retained the services of Alan Milstein, a New Jersey attorney who in recent years has represented many families in lawsuits involving problems that took place during clinical trials of new treatments. For example, he represented the father of Jesse Gelsinger, who died in a gene therapy experiment five years ago and got a rumored $2 million settlement in that case.

Milstein got a third of the reported $125, 000 settlement in the Quinn artificial heart case. His presence is a signal that the legal system that has made such a hash out of medical malpractice is heading full-tilt toward the world of human experimentation.

The problem with this lawsuit and other similar cases is that the lawsuits do absolutely nothing to help protect human subjects. Doctors still approach dying patients about their interest in trying new drugs or devices. Coerced by their disease, some patients and their families want hope, don't really read the informed consent forms, and sign on thinking a cure may result. Of course, cures almost never do result; early-stage research fails hundreds of times for every cure that is produced. Only later—despite all the informed consent forms, committee reviews, Food and Drug Administration analyses, and interventions by patient advocates—do the subjects and their families say that they did not really understand that experimentation is a messy, risky, miserable, and dangerous business in which people are sometimes harmed and even killed.

At that point, ambulance chasers like Milstein are their only option if they think that they have been wronged. But that is no option at all since, as is already clear from the malpractice crisis, the only guaranteed winners in lawsuits involving doctors and hospitals are lawyers.

For more than two decades, various committees and commissions have called for the creation of compensation funds for subjects injured in human subjects research. A no-fault system of subject compensation would help make it clear to families like the Quinns that research is not therapy. It would also prevent the kind of crisis now rampant in American medicine due to malpractice from spreading into research. If the federal government does not act, then those who simply don't grasp what is going on when a new drug or device is tested or who unfortunately are hurt or harmed will have only lawyers to turn to. If that trend continues, then it will be lawyers, not researchers, who ultimately decide what kind of research is going to get done in the United States. And if that happens, it is simply our fault for not making sure that those who sacrifice in human subjects research are treated fairly.

Commercial Concerns Should Take a Backseat to Public Awareness

LISTEN VERY carefully. Do you hear the sounds of ice cracking in the fiery domain where sinners are said to go upon their deaths? Has hell in fact frozen over? Can it be true? Has the American Medical Association (AMA) broken away from a long history of concern about protecting its members' self-interest and pushed forward an idea that is sound, bold, and in the public interest. Hell, yes!

The AMA has voted to ask the federal government to create a registry that would make publicly available the results of all drug experiments conducted on human beings. This means that anyone could look and see what drugs have worked, which ones are duds, and which ones are known to have possible dangerous side effects.

Academic researchers have a hard time getting research published if it does not show positive results. Negative studies don't get past most journal editors. Even if experiments that don't pan out do get published in the academic literature, you are not likely to hear about it. "Common cold still not cured" is a headline you are unlikely to ever read or hear.

More troubling than the problems academic researchers have getting negative results known is the behavior of private companies that sponsor studies. Pharmaceutical and biotechnology companies sponsor the overwhelming majority of late-stage tests of new drugs and medical devices. When private drug companies know that there are negative results or even injuries and deaths associated with their products, they are under no obligation to make that information known to you or the medical profession. The companies consider this data proprietary. They have lobbied to ensure that only the Food and Drug Administration is supposed to get this information, and even then some drug companies simply prematurely squash studies whose results they don't think they are going to like.

The most recent example of hiding negative results concerns children. Drug company studies showed evidence linking the use of antidepressant drugs to an increased risk of suicide in children. But they did not disclose that evidence. It was only when New York State's take-no-bull attorney general, Elliott Spitzer, starting hauling the offending drug companies into court that the damaging evidence was released.

The physicians of the AMA realized that without ready access to all experimental data—good, bad, and indifferent—they cannot know what is the best treatment for their patients. And they also know that without public access to negative data, all they can go on is what the marketing departments of the drug companies tout as the best drug to prescribe.

There is another reason to make sure that every bit of data produced in testing new drugs and devices on human beings is made public. It is part of the tacit contract that the researcher and sponsor of the study have with each and every subject.

If you agree to be in an experiment or to participate in a clinical trial to find new ways to treat your cancer, diabetes, Parkinsonism, asthma, depression, or migraines, you are told that the chance of directly benefiting from your participation in a study is at best unknown. Research simply often does not pan out. However, you are also told that if you choose to enter a study, then even if it does not work—and most studies do not—doctors will learn and future generations will benefit from that knowledge.

That is a great reason to participate in medical research. But if negative results are not published, if bad outcomes or deaths are swept under the rug, then nothing is learned. And the promise that was made to those who put up with the risk, inconvenience, and hassle of research is broken.

Medicine will be much, much better off if commercial concerns take a backseat to the need to know. All data from every experiment, regardless of who pays for it, should be kept in a national registry that is accessible to anyone who wants to see it—not just because such a registry will make medicine more effective, cheaper, and safer but also because it is what every subject in a medical experiment has the right to expect.

Research Ban at Hopkins a Sign of Ethical Crisis

THE SUSPENSION of most federally funded research at Johns Hopkins University medical center, albeit only for a few days, is a mind-blowing, incredible event. It is more staggering than any medical news or breakthrough this year. To see research suspended for ethical violations at what is arguably America's finest medical institution is almost beyond belief.

Johns Hopkins, the leading recipient of government research money in the United States, has been banned from nearly all federally supported medical research involving human participants—the death penalty for any medical institution.

A penalty like this is almost never imposed since the impact is incredible: subjects cannot complete their studies; patients cannot enroll in new trials; findings in time-sensitive studies are not reported. The money is not there to pay the technicians. Researchers cannot pay for their lab equipment and materials. Scientists are at risk of losing months, if not years, of work.

Some will argue that Hopkins is the culprit. After all, that is where healthy Ellen Roche participated in a research experiment that was designed to assess her response to a substance that was supposed to provoke a mild asthma attack in order to better understand how the body fights the lung disorder. Instead, the experimental treatment wound up killing her.

Hopkins has already admitted that it did not do what moral researchers must do in such experiments—obtain an ironclad informed consent and have an in-house committee carefully monitor all experiments that are done simply to obtain knowledge and not to benefit the subject.

But the suspension of clinical research at Hopkins is a symptom of a much deeper disease—the collapse of adequate protections for those involved in research at every American medical center, clinic, testing facility, and hospital. And if a culprit is to be singled out, it is that disease, not one institution.

The system for protecting human subjects research is not simply sick—it is dead. The roll call of institutions with ethical violations and problems over the past five years has become a "who's who" of research—the Fred Hutchinson Cancer Center in Seattle, the University of Pennsylvania, Duke University, the University of Illinois at Chicago Circle, the West Los Angeles Veterans Administration Hospital, Virginia Commonwealth University, the University of California at Los Angeles, the University of California at Irvine, and others. The list goes on and on and on.

When one of the leading medical schools in the world has all research suspended for noncompliance with federal rules governing the use of human subjects, then it is time to pronounce the patient dead and mount a systematic effort to find a cure.

America needs a complete and thorough reexamination of our current laws and policies designed to protect human subjects as well as their implementation, the adequacy of educational efforts to teach researchers their duties and obligations, and an honest assessment of the amount we are spending to promulgate and enforce these protections.

The administration and Congress are not taking the protection of human subjects seriously enough.

Medical research is not only a crucial factor for the health and well-being of all Americans and their descendents but also a major engine in our economy and a crucial source of innovation in every aspect of human life around the globe.

To say that subjects are not safe at an institution with the preeminence of Johns Hopkins is not an indictment of that institution. It is an indictment of our societal failure to attend seriously to the crisis that has been building for years all around the nation as school after school and company after company are found guilty of ethical violations when it comes to their clinical research.

Research carries risks. To show that a new treatment works, research must sometimes be done by comparing that treatment's effectiveness with placebos, inactive substances that we know won't benefit the patient's disease to the same degree as an active drug.

Research must sometimes be done using drugs that provoke adverse reactions and symptoms in subjects. Research must be done on the healthy as well as the unhealthy. Research must be done on children, and the mentally ill, and the cognitively impaired, and the drug addicted. Research must be done in the emergency room. Research must be done on the dying.

Research must be done, but it must be done ethically.

It is time for the Department of Health and Human Services, the White House, and Congress to address the issues, appropriate the requisite funds and resources, and set about fixing this problem. Those who altruistically give of themselves in human research deserve nothing less.

Johns Hopkins is not the problem. But the calamity that has temporarily paralyzed that proud institution should be the start of a solution.

Research on the Newly Dead

THE UNIVERSITY OF Pittsburgh is proposing to undertake research that is sure to raise a few eyebrows. Scientists and doctors at Pitt are enacting a policy that would let them conduct experiments on the newly dead. In particular, someone who died while on life support might be used to test a genetically engineered pig heart or to try out a newly developed artificial liver.

Why study the dead? Well, to put it simply, they have much they can teach us.

By putting, say, an organ from a pig into a newly dead body, it is possible to learn whether it will be rejected right away without having to compromise the health of a living human being. And studies of the dead might allow us to try out doses of drugs or viruses that would never be accepted as ethical to give to any human subject. The dead might even be a place to learn about the effects of pesticides or biological weapons on the human body.

A tiny amount of research has been done on the dead. Many years ago, an early version of the artificial heart was tested on brain-dead subjects at Temple University Medical Center. But researchers and medical centers have been too leery of the bad publicity likely to surround an announcement that the dead are being used in medical research.

This research is important. It should get done. And it is ethical to do it if care is taken to make sure that the person who has died has given voluntary consent to serve as a subject after death. People today consent to organ donation and to donate their bodies for dissection. Some would undoubtedly let themselves be studied when they die.

The other crucial ethical safeguard is that there have to be clear limits to how long the dead can be experimental subjects. Families and friends need to grieve and come to terms with death, so bodies cannot be kept for long even if the purpose, research, is ethical.

These details can be worked out. Pittsburgh is right to try to shape a policy to permit research on the dead. To not learn from the dead is to put the living at too great a risk.

Will We Ever Debunk Our Mythology about Human Subjects Research?

WHY IS IT that every time you turn on the radio, there seems to be another scandal involving medical research? The Bush administration has just announced a nationwide review of all medical research at every veterans' hospital in the country. This review was triggered when it was found that no informed consent had been obtained in research involving vets in Fargo, North Dakota, and Albany, New York. The Department of Veterans' Affairs said that it had also found serious noncompliance with federal rules at six other VA hospitals, including the falsification of research data and the failure to adequately and accurately disclose the risks of participation to prospective subjects.

Research scandals are not confined to the Veterans' Administration. There have been an incredible string of violations, cover-ups, and problems at academic medical centers, private hospitals, and corporations over the past few years.

The standard response of Congress and the administration to each scandal is to call for stricter adherence to informed consent. Unfortunately, informed consent does not work. But to say this is to challenge one of the strongest myths in American culture—that each person is capable of self-determination. But we need to get past this myth if we are ever going to bring an end to scandals in medical research.

Why is informed consent a myth? Well, there are a slew of subjects—fetuses, children, the severely mentally disabled, people who are in comas—who could not possibly give it. But I am not talking about them. I am talking about the average person.

Think about it—can ordinary folks really give informed consent? Some do not speak English well. Some are not well educated. Some are intimidated by anyone who speaks to them wearing a white coat. Some are so eager to find hope from gaining access to research that they will do anything a doctor asks. Experts in mental health tell us that close to 8 percent of all Americans have an undiagnosed or untreated serious mental illness, but no one is ever deemed too incompetent to be in research by a researcher. And for some, the lure of compensation ends the possibility of careful deliberation about risk and benefit. Huge numbers of Americans cannot do what our mythology tells us they can—weigh the risks and benefits and decide whether to participate in research.

These facts are borne out in the various studies that have been undertaken of informed consent. All potential subjects have a hard time processing technical information that appears in informed consent forms. They do not recall what is said in ten-, fifteen-, or twenty-page forms.

Yet because the dominant value of American society is respect for self-determination, we continue to rely on a protection—informed consent—that often fails those it is intended to serve. And that is when those soliciting consent try their hardest to obtain it. Matters are worse when, as in the case of the VA scandal, researchers don't try at all.

If human subjects research protection is to really improve, then we must design a system that is not based on mythology. Sadly, there is no evidence that the research community, patient advocacy groups, or legislators are inclined to give up the myth. Until we rely on something more than just informed consent to protect subjects, the scandals will not go away anytime soon.

Part VI

Health Reform

Cause Célèbre

A **GROWING NUMBER** of celebrities are using their star power to raise funds and awareness for an array of diseases. Julia Roberts is pushing for federal dollars to fight Rhett's syndrome. Supermodel Christy Turlington raises awareness on emphysema. *West Wing* star Brad Whitford is the voice for autism. And, of course, Jerry Lewis has been hosting his annual Labor Day telethon for muscular dystrophy for nearly forty years. But do celebrities on soapboxes really help?

Lewis has his critics. Some people wonder if he is exploiting the parade of "Jerry's kids" who appear on the show along with all the singers, plate twirlers, and jugglers. But his telethon has, after all, raised $1.7 billion.

And Roberts went before Congress to push for funding for a very rare genetic condition called Rhett's syndrome. This is a terrible affliction but not one that affects many people.

She was following Christopher Reeve, who had just been there asking for money to find a cure for paralysis. And Michael J. Fox is well known for his support for research on Parkinson's disease. There are plenty more celebrities plying the halls of Congress or leading fundraising campaigns looking for dollars to be directed toward the disease they care about the most, from AIDS to breast cancer to ovarian cancer.

And they should all be applauded. I am happy that Lewis still takes the time to do the telethon. Roberts deserves praise for bothering to go before Congress. And Fox has shown nothing but commendable caution in the way he has tried to draw attention to a terrible disease without letting his celebrity overwhelm the exciting science that may lead to a cure in the not-so-distant future. They are most certainly doing the right thing.

The problem is simply that there are not enough celebrities doing what Lewis, Roberts, and Fox do. Some diseases, such as alpha-1 antrypsin disease, Canavan disease, bulimia, or lupus, have no celebrities willing to go to the mat for them. Some ailments are just too stigmatized or uncool to attract celebrity support. It is hard to imagine J-Lo or Jennifer Aniston leading a march on Washington to demand more research on urinary incontinence.

The problem with celebrity fund-raising is simply that it is not fair. Celebrities who try to lobby Congress sometimes don't know the science well enough to know what is the best way to spend the nation's research budget. So the budget can get distorted, and some people with real diseases that have a real shot at a cure if only the money were spent on them lose out.

Clearly, it is an advantage to have celebrities involved in drawing attention to diseases and raising funds.

But remember, for every celebrity who tells us about the hope for a cure or the need for Congress to act to win the war on some malady, there's probably another group that is just as worthy but that has not been lucky enough to attract a Jerry Lewis.

Cheap Drugs Are Not the Answer to the African AIDS Crisis—Better Infrastructure Is

F DRUGS ARE credited with transforming HIV from an almost certain death sentence to a chronic, manageable disease in developed nations, then isn't it our moral duty to make sure that people in the poorest countries hardest hit by the AIDS epidemic have access to these same medications? The answer to that question may seem obvious. But it isn't.

AIDS is cutting a terrible swath of death through sub-Saharan Africa and parts of Asia. It is the leading killer of people under forty in many African countries. And AIDS threatens to explode with devastating consequences in Russia, India, and parts of Eastern Europe.

AIDS is still a major problem in the United States. But it is not the lethal plague it once was. What accounts for the difference in the body count between Africa and America? The answer is drugs.

The lethal sting of HIV has been somewhat blunted by a new generation of powerful drugs that, while not a cure for AIDS, do slow the effects of the virus and allow people to live for many years.

These drugs cost a lot of money: $15,000 per person per year is the figure most often cited. Obviously, a person living in poverty—as is often the case in areas of the world hardest hit by AIDS—cannot afford them.

So the solution seems equally obvious: Make the drug companies that produce the requisite drugs—and make a lot of money selling them—offer them at deeply discounted prices or give them away for free.

In fact, international agencies, governments, and patients' rights groups are placing enormous pressure, with some success, on the huge multinational drug companies to do exactly this. The companies are being told that they must either give anti-AIDS medicines away to poor nations such as South Africa or allow smaller companies to infringe their patents and produce cheap, generic copies at greatly reduced prices.

Giveaways sound good. Huge pharmaceutical companies make billions each year in profits. So why not force them to help the poor, who will otherwise die?

Because, in the end, giveaways are a bad idea. While drugs are an answer to the AIDS plague in North America and Western Europe, they are not the solution for Africa and many other extremely poor nations.

The reasons are simple. Drugs designed for people in more developed countries will not work as well for people living in countries that have no hospitals, clinics, clean water, sewers, roads, or doctors. Some of these drugs must be taken with food to work effectively, for example. So in nations on the edge of famine, they will not do much good.

It also is not easy to stay on the latest generation of antiviral medications. A person infected with HIV must take more than thirty pills every day. Some medicines have to be taken with water, but without food. Some must be ingested early in the morning, others late at night. Probably ten or more different prescriptions are involved in the life-preserving regimen that is slowly changing AIDS from a lethal condition to a chronic one in developed nations.

So what happens when a person can't get to a clinic to fill all these prescriptions? Or fails to take all his medication as directed? Nastier, drug-resistant forms of HIV that are even harder, if not impossible, to treat emerge and proliferate.

Drugs also don't get to the heart of the AIDS epidemic. Simply throwing drugs at countries that have no educational or public health programs will not slow the spread of the disease. Unless drug distribution is linked to public health campaigns, AIDS will not be stopped. And the high cost of drugs means that eventually, the patience of drug companies and their shareholders for giving the medications away will run out, and the media will get bored with the story of people in faraway lands dying in droves for want of medications. After a few years, charity will disappear, only to be replaced by the same old indifference.

What these poor nations need is what they always have needed: a solid health care infrastructure. Those dying of AIDS need drugs, but they need a clean bed to lie in and safe water to drink even more.

No pill should be given away to any poor nation without a commitment to build a clinic, a road, a source of clean water, or a public health program aimed at AIDS prevention.

Drugs alone are not the answer. The world's poor who have AIDS needs clean water to take their pills.

Humility or Hubris?

I **HEAR IT** a lot—people are always grabbing me at cocktail parties or water coolers in order to complain, "I wish doctors were not so arrogant."

This is not a sentiment that people express about their particular doctor. Most people actually like their personal doctor. But, in the abstract, when Americans are making up their list of who it is out there who has grown insufferably enormous for their collective britches, doctors rank right up there.

As a person who works in the ethics of health care, I can say that there is very little I can do to provide an antidote to medical arrogance. Exposure to ethics does not make you a nicer or more humble person. If it did, then meetings of philosophers and theologians would not involve such prolonged exposure to many preening and self-important souls. But those who worried about overbearing and pompous physicians need worry no more. Lowly viruses have proved to be quite capable of deflating the collective ego of those wielding the stethoscopes and the white coats.

As the New York area's recent experience with West Nile virus–bearing mosquitoes should remind us all, there are plenty of nasty critters out there who can stump and then defy even the greatest medical minds. It took forever to figure out exactly what was going on when this virus first appeared in the United States. Not a disease that would give any physician grounds for haughtiness.

The AIDS virus continues to prove incredibly resilient to our most powerful drugs. Although we are keeping those infected with HIV alive longer, we still do not have a drug that cures. And vaccines have proved incredibly difficult to perfect.

Hepatitis C virus has nothing but contempt for the brilliance and skill of modern medicine. While infection rates have fallen, it remains the case that this decades-old plague still has treatments available that work at most 30 percent of the time.

Mad cow disease not only remains undetectable but is completely incurable. Alzheimer's disease is yet another plague with some sort of microscopic process at work that is opaque to the brightest of medical minds.

And the reminders to be humble are not confined to microbes that try to use us as potting soil. The tragic death of a young man in a gene therapy experiment at the University of Pennsylvania is a sharp reminder that when we try to use viruses for our purposes—in this case, to deliver a gene missing in the livers of some human beings—it is sometimes the virus that has the last word on what is going to happen.

Some years ago, an expert in infectious disease rather arrogantly opined a public lecture that, with the demise of tuberculosis and the conquest of polio, medicine would have no more infectious diseases to do battle with. Well, TB is still here. And shortly after the war on infectious disease was declared over, HIV exploded around the world. Nature is making it very clear that lurking out in our world are still more mutating viruses and pestilent bacteria that are more than capable of making a mockery of any declaration of final victory in the war on microbial diseases.

Medicine has to learn an important lesson. The lowliest organisms can confound even the mightiest medical experts. Whatever medicine's achievements—and there are many—humility is a better guide to dealing with the dangers and risks in the world around us than is hubris.

Fiddling While the Health System Burns

JUST HOW BAD is the state of health care in America? Well, consider two recent developments that shine a spotlight on a system that was already showing signs of severe distress even before the Supreme Court decided to let HMOs off the legal hook. In Colorado, the rich are paying what amounts to bribes to make sure that they are at the head of the line when it comes to getting health care; and in Tennessee, the poor are basically being told to get lost.

Denver was the setting a few weeks ago where more than one hundred physicians from around the United States attended the first meeting of the American Society of Concierge Medicine. Concierge medicine is a special, high-end form of medical care that guarantees that if you need treatment, you will get it, without a hassle, seven days a week—but only for an extra fee. If you can pay amounts that range from $20 to thousands of dollars a month, you can guarantee that your phone calls will be promptly returned by your doctor and that you'll get special attention whenever you're admitted to a hospital.

Now, one might wonder why it is necessary to pay a bounty to get a doctor to call you back, especially if you are already paying through the nose to belong to a managed care plan. The answer is that under the watchful eye of managed care and insurance companies, the quality of care has gotten so awful that doctors sneeringly refer to it as "hamster care." Only those patients who pay more are going to get treated by the "concierge" doctors who get off the daily treadmill and practice good medicine—providing the sort of attention and service that our parents and grandparents took for granted.

Think that giving the rich special access to health care is unfair? Consider what is going on at the same time in Tennessee.

Tennessee is making over its state Medicaid program known as TennCare. If this program gets implemented, many of the poor elderly, children, and disabled in Tennessee who rely on Medicaid will be told simply to get lost. And other hard-pressed states may well follow suit.

Governor Phil Bredesen, a former HMO entrepreneur, sees the challenge of health care for the poor in Tennessee in very stark terms. In a speech last February, the governor described the state Medicaid program as nothing more than an open checkbook that is continuously being raided by "doctors and hospitals and advocates" who "decide what is needed." Well, who should be deciding what is needed for medical treatment if not doctors and hospitals and advocates? Not under TennCare, if the governor gets his way. Bureaucrats, not doctors, will pick how the poor get treated.

Historically, decisions about what drugs or treatments a patient received were chosen by a standard of care known as "medical necessity." Doctors determined what was medically necessary based on local standards of medical practice, and if they did not practice according to this standard, they could be found guilty of malpractice. TennCare does away with the established standard and replaces it with a new one—"adequate care." If a bureaucrat in the Tennessee Department of Health thinks a low-cost drug or treatment or even no treatment at all is "adequate," then that is what TennCare will provide.

Under the new definition, preventive care and many pain medications will no longer be funded. And no drugs at all will be available to treat poor kids with life-threatening conditions such as cystic fibrosis, cancer, or asthma and no antihistamines or gastric acid reducers for anyone of any age. If you want to protest these inadequacies, you might be able to find a doctor willing to plead your case to a special state-established foundation.

No one wants to see any state dissolve in a sea of red ink. But how can any American stomach a public health care system that is so unfair to people who aren't rich? Surely there are less drastic steps that a state like Tennessee could take that would let doctors decide what is appropriate care for children, the disabled, the chronically ill, pregnant women, and the elderly poor.

Those now seeking public office must come up with something better than medicine prescribed by bureaucrats. Whatever solutions they arrive at, doctors, not bureaucrats, must be in charge of our health. That is something to remember when we enter the voting booth.

No Coverage for Kids a Moral Failure

EVERY ONCE in a while, the president, the head of the American Medical Association, or the surgeon general will stride up to a microphone and proclaim that we have the finest health care system in the world. While not quite as bizarre as Howard Dean's now-infamous scream at the wake held after his defeat in New Hampshire, such statements are pretty loony. Why do I say this? Because we have children in the United States who lack health insurance and thus are not getting the care they need. And the problem is getting worse.

The state of Alabama put a freeze on enrollment into the state's ALL Kids insurance program. So every child in Alabama, no matter how serious or life threatening their ailments, now is on a waiting list rather than in a doctor's office. So, if you are a young person like six-year-old Stephanie Romero of Brewton, Alabama, and you have rheumatoid arthritis so severe that you cannot get in and out of bed without your mother's help, well, too bad for you. Not only do you have no access to the finest health care system in the world; you also cannot even get the basic care you would get if you lived in any other economically developed country on Earth.

Alabama is hardly alone in mistreating its children. The state of Georgia lopped 12,500 women out of eligibility for prenatal care. Tennessee is denying basic supplies for the treatment of diabetes to kids. Tens of thousands of kids remain without insurance in the states of Florida, Arizona, and North Dakota. Get the picture?

There is no way you and I live in the country with the finest health care system in the world. Any nation as rich as this one that has even one child who lacks access to health care because they have no insurance is a nation that is a moral failure.

The so-called finest health care system in the world has nearly ten million uninsured children in it. President Bush is doing nothing to fix this problem. Governors are balancing their state's economic woes on the backs of their youngest citizens. Health insurance companies are doing little to solve the problem. And the problem just keeps getting worse. Shouldn't anyone who aspires to or holds national office tell us what they propose to do to make sure that children get the health care they need?

New World Calls for New Health Care

GOVERNMENT SHOULD provide a safety net for all Americans. Everything is different now. This is a mantra that has echoed all over our nation since the terrorist attacks. But one thing has stayed the same—our lousy health care system. And now, as we cope with the loss of jobs as well as lives, it's more important than ever for the government to provide a safety net for all Americans.

We once lived in a world so secure that our only sense of daily risk came from the titillation provided by reality television programs and Hollywood thrillers. That world is gone. We once lived in a world where our greatest fears were stoked by images of human clones and computers run amok. That world is gone. We once lived in a world where we had grown callous and indifferent toward the tribulations of our neighbors. That world is gone, too.

I have been thinking a lot about the world we left behind lately while listening to the words of the spouses, friends, lovers, and children of those killed in the terrorist attacks. One woman in particular caught my attention. I did not catch her name on the television interview, but her husband was killed in one of the World Trade Center buildings. She has five children and medical and psychological health care bills to pay. She is not sure what she is going to do about paying them. Everything is different now. Except we still have the same pathetic, immoral health care system we had the day before the terrorist assaults.

This mom and the thousands more who are dealing with the same problem of how to pay their health care bills should not face this problem. All the rest of us who are now collectively soldiers in the war on terrorism should not have to worry about health care bills. It is time to do what is right and make sure that the old broken world of health care insurance is gone, too.

It may seem crazy to call for a massive push to create a new social program at a time when the economy is reeling and the nation is at war. But strange as it may seem, this is precisely the time to make sure that no

American need wonder how they are going to pay for their medical bills or those of their kids.

The new breed of war knows no boundaries. It does not distinguish between soldier and housewife. It can cost you your job and your health insurance overnight. It can come right to your doorstep and leave you hurt or disabled.

In the past, efforts to create a national health care plan have foundered when private interests have defeated the public good. Americans can no longer afford to put the public good behind private interest. No mother should be worrying about how to pay for her kids' medical bills because her husband has been killed by terrorists.

President Bush and Congress ought to announce a plan to mandate that every American will have access to high-quality health care. The plan should not put the government in charge of providing care. It should simply state that every American will have access to care— through privately purchased health care plans, coverage earned through military service or employment, or government-sponsored insurance vouchers. What we have done for our automobiles—universally insure them—we must now do for our neighbors.

After the World Wars, Korea, Vietnam, and the Gulf War, those who served knew that they would get the medical care they needed. A grateful nation promised them what they had earned. In the new world where each of us is a target and every American is a veteran, we must make the same promise to one another. Every American must know that regardless of what terrorists do, there is a safety net for them and their families.

Our Dying Health Care System

AMERICAN HEALTH policy is a moral disgrace. The major issue that has gripped the current administration and Congress is whether to extend prescription drug coverage for the elderly. Now I yield to no one in my respect for the elderly. But with all due respect, enhancing drug coverage for the elderly is not the nation's number one health policy challenge. It is not even number two or three. Or even four!

The number one problem is that our health care system is so completely broken that it does not cover a big chunk of Americans. Close to forty-two million Americans have no health insurance at all. And the number of uninsured is growing as managed care plans continue to pull out of the Medicare system and more and more doctors refuse to accept any elderly patients at all given the lousy level of reimbursement they receive. Add to this that most Americans have coverage for long-term care, and it is easy to see what the big problem is—lack of insurance and thus access to quality care.

Number two is that the bad economy is destroying the Medicaid safety net for the poor. States are hacking Medicaid to death.

In Oregon, there no longer is, for all intents and purposes, health care for the poor. The mentally ill are being told to fend for themselves. Those who are schizophrenic, severely depressed, or autistic will, Oregon officials say, just have to make do. This is immorality at its most excruciatingly awful.

The third problem is that costs continue to rise. The price of prescription drugs is going through the roof. Employers' health insurance premiums have multiplied twelve times over the last decade. Those who get insurance through their jobs—and that is a whole lot of Americans—are paying higher and higher copayments and deductibles. Many retirees are finding out that their health benefits are simply gone—the companies that promised them are defunct or bankrupt.

Fourth but not least, malpractice is sucking the life out of medicine. It is impossible to overestimate the anger levels that are present among physicians about exploding insurance premiums. If this system were really bringing justice to those who are injured or reducing culpable errors, then maybe there would be some excuse for the exploding costs. But it is not, and everyone knows it, including the trial lawyers who are getting rich from the status quo.

Sure, the elderly face problems in getting access to many of the drugs they need. But on the scale of moral outrage, that does not compare with 20 percent of the population unable even to get into a doctor's office. Or those who are severely mentally ill going without any medical treatment at all.

Politicians are upset that the elderly don't have the coverage they deserve for medicines. It is a moral scandal that they appear not to give a damn about fixing a health care system in which so many lack coverage for anything.

The Moral Tragedy of Chronic Illness

A **FEW WEEKS AGO,** I was presented with the following case: A young woman, sixteen years of age, had been brought to a psychiatric facility in Philadelphia. She suffered from anorexia. The staff of the hospital called to ask a very simple and straightforward question: was there any end to their obligation to care for this young woman?

This simple question presents some of the most profound ethical questions raised when an illness is chronic. And it also reveals some of the uniquely problematic and disturbing aspects of America's utter failure to deal with the realities of chronic illness.

The young girl—let's call her Loren—had suffered from anorexia since the age of thirteen. She had become very concerned about her body image about the time she began to undergo puberty. She was obsessed about her weight. She found herself very satisfied by feeling in control of her body. Bulimia and dieting became a way of life for her.

Her parents, who had divorced when Loren was ten, became concerned about her obvious loss of weight shortly after her thirteenth birthday. They had read about anorexia in magazines and seen a television program on the subject, so they were quite terrified as their daughter began to grow thinner and thinner. Each parent tried to talk with Loren about the weight loss, but she would simply withdraw or deny that anything was wrong.

The issue of weight loss became an issue between the parents as each one struggled to try to respond to their daughter's behavior. Each blamed the other for placing too much emphasis on food and fashion and for pressuring their daughter in a variety of other ways.

Eventually, as Loren's weight dipped well below one hundred pounds on her five-foot, five-inch frame, and as her friends and teachers began to make nervous comments, the parents decided to look into getting some form of treatment for Loren. What they learned amazed them.

The mom had no real health insurance from her job. The dad did, but it had very limited benefits for outpatient, nonhospital services. The only way Loren would receive any care from a mental health professional was if she needed to be institutionalized in a hospital because her loss of weight was directly threatening her life. There were outpatient clinics and counseling for kids like Loren, but these cost hundreds of dollars per hour and were well beyond their means.

The lack of coverage by any insurance programs meant that an adolescent girl with a serious disease could receive care only if the condition degenerated to the point where her life was in peril. Loren did have a few sessions with her high school psychologist, who quickly realized that Loren's anorexia was beyond her capacity to treat. So Loren wound up bouncing in and out of the emergency room at a number of psychiatric hospitals.

When her weight got so low that her doctors thought she might die, she would be admitted for a three- or four-day stay. Her insurance carrier was always on the phone, quick to demand that she be discharged since her insurance only covered required acute care services, and anorexia was not in itself a basis for treatment.

Nothing was done to get at the underlying core of Loren's illness. Chronic mental illness is simply not covered by most insurance plans. This meant that Loren had to be in crisis to gain entry to health care. Not only is this approach inefficient, but it had the effect of turning the health care providers against this young woman because the only time they saw her was when she was so emaciated that she was almost dead. They felt they could do nothing for her to get at her disease, and having her appear periodically in their intensive care unit was frustrating to them beyond their endurance.

In one sense, the answer to the medical team is clear. They must continue to treat Loren. They would not even think of turning away a heart attack patient no matter how many times that patient appeared in the emergency room. But chronic illness is not seen in the same light. Chronic illness stigmatizes those who have it. It makes them seem noncompliant and uncooperative. And the acute health care system where Americans put the bulk of their health care dollars has neither the orientation nor the services to cope with the problems of a Loren. The stigma of chronic illness only serves to reinforce the fact that a person with a mental illness like Loren is kept from gaining access to the services she needs.

If Loren needed a life-saving transplant, no matter how long the odds, no matter what the cost, she would get it. Americans do not say no to sixteen-year-old girls who are dying. But if Loren has a disease that kills her slowly, that stigmatizes her as crazy, that is frustrating to treat and lacks the glamour of having a high-tech solution, then it is all too easy to say no to her and the tens of millions of other Americans with chronic illnesses. But that is what we are doing in American health care today. And that is a moral tragedy.

Loren's case is just one of thousands of similar situations faced by patients, family members, physicians, and social workers every year in America. Only by understanding the challenges faced every day by people with chronic illness can we do a better job of tailoring our health care system to meet the needs of those with chronic illness and to prevent future tragedies like Loren's from occurring.

Part VII

Human Cloning and Stem Cell Research

Cloning: Separating the Science from the Fiction

AN ALLIANCE OF abortion opponents, social conservatives, and biotechnology-phobes wants you to believe that human cloning is always unethical, even when it's done for the purpose of finding cures for horrible diseases. It isn't. Understanding the reasons why is very important since the quest to crush cloning is likely to be renewed later this year.

There is so much confusion and misunderstanding surrounding cloning in part because it's tied to our ongoing battle over abortion. Abortion opponents continue to look for any opportunity to secure legal recognition for the personhood of an embryo. When abortion foes like President Bush call for a ban on cloning, they are really using cloning as a tool to try to pry open *Roe v. Wade*. By claiming that cloned embryos are people and that their destruction has to be outlawed, they hope to get legal standing for all embryos. A ban on all forms of cloning would lead to bans on the destruction of all human embryos, cloned or otherwise. That would likely spell the end of abortion as well as in vitro fertilization and most forms of prenatal genetic testing in the United States.

The view that a cloned embryo is a person is, however, wrong. There is a huge difference between a cloned cluster of embryonic cells in a petri dish that could yield disease cures and a baby.

Cloning Myths

The future of cloning has far more to do with cells than with people. The push for a total ban on cloning rests on several myths and far-fetched scenarios that have gained way too much currency in Congress, the Oval office, and the media.

MYTH 1: Cloning for scientific research and cures, known as therapeutic cloning, sends humanity hurtling down the slippery slope toward the inevitable cloning of human beings.

The best rebuttals to this argument are the thousands of failed attempts to clone animals. Cloning has a terrible track record in making embryos that develop into fetuses, let alone make it to birth.

Attempts to develop cow embryos into animals fail 85 percent of the time, while more than one-third of those clones born alive suffer serious life-threatening health problems. Despite a lot of effort, no one has managed to clone an adult monkey or any other primate. Nearly all experts on primate cloning believe that monkeys and human beings will never be cloned because the biology of primate reproduction is simply unlike that of cats, goats, sheep, and mice.

At present, cloned human beings exist only in science fiction, lurid tabloids, and the boastful and bogus claims of sham scientists and cult kooks. For now, the only destiny for cloned human cells is to help scientists understand and cure diseases.

MYTH 2: The pursuit of therapeutic cloning will lead to the exploitation of women for their eggs, since billions of eggs will be needed.

The number off eggs that is needed is grossly exaggerated. To go forward, cloned embryonic stem cell research would need thousands of eggs, not billions.

Women throughout the country are already providing thousands of eggs to infertile couples, and more eggs could be donated by women who just want to help scientists find cures to diseases or disabilities or who simply want to help find cures for their own diseases.

MYTH 3: What if someday scientists find a way to clone humans safely? Unscrupulous people of means will try to crank out armies of Hitlers or lines of designer babies.

Despite what Hollywood has to say, Hitler is not coming back, even if we could clone his genes. Genes influence but do not determine personality or behavior. Rather, each of us is shaped by the unique time, place, environment, and circumstances in which we live. To get a Hitler or a Saddam Hussein or any other tyrant, you need more than their genes; you need their mothers, their fathers, and the environment in which they grew up. Science will never be able to clone particular people to order.

If human cloning ever "worked"—which is highly improbable on a mass scale—it could not bring back the dead, create a new pathway to immortality, or furnish the means to create new strains of dictators.

MYTH 4: There are other techniques for finding cures, including the use of adult stem cells, so there is no need for cloning.

The reason to clone embryos is that the resulting cells and tissues will have the same genetic makeup as the person they come from. Therefore, they can be transplanted back into the person without fear of rejection.

The reason that adult stem cells do not offer an equally valuable alternative is that embryonic cells are the only cells capable of turning into all the various types of cells that are needed to fight disease, disability, and death. And no one has figured out yet how to get adult stem cells to revert back into this omnipotent state.

If your child is dying, you want all research avenues pursued, and that includes both embryonic and adult stem cell research.

The Bottom Line

The bottom line is that cloning for cures has the potential to do enormous good by saving the lives of millions of people and ending agony for millions more. These human beings and their loved ones aren't interested in pieties and abstractions and science fiction. They are desperately seeking help for their ailments, and they need to have medical scientists free to pursue those answers and cures. Banning all human cloning would be a highly unethical thing to do.

Rather than shackle American scientists, the U.S. government should encourage cloning research. The needs of children confined to wheelchairs, of parents dependent on oxygen tanks to breathe, and of friends imprisoned by the creeping paralysis of Parkinson's far outweigh the moral status of cloned cells that will never leave the petri dish. Myths should not be the basis for public policy when cures hang in the balance.

Cloning Flicks Offer a Moral Lesson

HOLLYWOOD WANTS us to be afraid, very afraid, of human cloning. From Sylvester Stallone's soporific *Judge Dredd* to Arnold Schwarzenegger's hilariously stupid *The Sixth Day*, the denizens of Tinseltown remain firm in their belief that they know how to scare the average denizen of the local multiplex.

Hollywood has made cloning into a bogeyman despite the fact that there is no way anyone will make a human clone anytime soon. Even the scientists who are best at cloning animals can make it work only about 1 percent of the time. There is no reason to think that human beings will be any easier to make and every biological reason to suspect they will be a lot harder to do.

Not to be deterred, however, Hollywood portrays a future world gone haywire as a result of businessmen—driven by a lust for money—selling the public the products of amoral, egomaniacal geniuses.

The problem with these recent theatric examinations of cloning is that (1) they stink, and (2) no one is in the least bit frightened by any of the philosophical nonsense that interrupts the machine gunning and muscle flexing. (Well, OK, I confess I was a little frightened at seeing Arnold, but it was only from watching him agonize as he sought to voice his insights about the ethics of human cloning in response to interviews by Barbara Walters and Katie Couric on two national television programs.)

Perhaps we remain unmoved by films about human cloning because the nightmare is not our nightmare. The horror that can happen when greed and renegade genius mix is more apt in describing what Hollywood brings to the screens in our malls than it is at depicting the consequences of cloning.

There are movies that manage to scare us and do well at the box office as well. But they are mainly about what happens when something is done to clone animals or plants or to alter their biology. The best of the genre of bioterrorflix—*Jurassic Park*, various versions of *The Fly*, and such 1950s classics as *Them*—involve nature gone nuts after we get done screwing around with its denizens.

I think there is a moral lesson nested in all this celluloid bioethics. We are much more worried about what we will do to nature with biotechnology and science than we are about what we will do to ourselves. We are actually, if subtly, used to changing ourselves. We do it by means of medicine, agriculture, and engineering from the moment we are born. Genetic change is change that is staggering in what it might lead to, but it is still nonetheless the kind of change that we already do to ourselves.

But ruining nature is different. The animals and plants have done nothing to bother us, so bothering them seems unjust. And if things go wrong with our interventions in nature—think killer bees, disease-resistant bacteria, extinct species, Starlink corn, dead Monarch butter-flies—then we, as movies about what happens when you try to fool Mother Nature nicely illustrate, most certainly will pay the price for our arrogance.

Pundits, bioethicists, and columnists like to think that it is human cloning and human genetic engineering that will stir the greatest moral controversies of the twenty-first century. Perhaps not. If the cinematic arts are any real indication, Arnold just makes us laugh, but hurting a dinosaur via genetic engineering is no laughing matter. We are very worried that dinosaurs, other animals, and plants will find a way to bite back.

Perhaps it is no accident that there have been few demonstrations against human genetic engineering but a large number against geneti-cally modified food.

Embryonic Cloning Feat
Points to Problems
with Bush Policy

THE CLONING OF a human embryo has created an entire new source of those miraculous stem cells that hold promise for treating a wide variety of currently untreatable diseases—and what is perhaps a more ethical source. The feat also points to the inadequacies of the Bush administration's policy on funding cloning and stem cell research—a policy that holds the American public its ultimate victim.

The biotech company Advanced Cell Technology announced that it has created two kinds of human embryos. One was made using traditional Dolly-style cloning techniques. The other involved tricking a human egg into starting to develop on its own, without benefit of a sperm.

The Worcester, Massachusetts–based company says its goal is not to make a baby or a human being but to find ways to create new sources of stem cells—master cells that have the potential to turn into any type of cell in the body.

The idea is to use a person's own DNA to make an embryo or, in the case of a woman, her own egg to make an embryo. Stem cells could then be extracted from these embryos and used to grow whatever cells or tissues a person might need—even whole organs for transplantation.

This strategy would obviate the rejection problems that haunt transplants today, as an embryo created from the recipient's own DNA would be the source of the replacement organs.

The only other two ways to get embryonic stem cells involve making human embryos by mixing sperm and egg in a petri dish and then destroying them, or using "spare" human embryos that have been left behind at infertility clinics by couples that no longer want to use them to try to have babies.

The prospect of using embryos as a source of stem cells set off an enormous debate that was only quieted when the president said that the federal government would continue to finance stem cell research—

but only if stem cells produced before August 9, 2001, were used in the work.

The problem with the president's policy is that it is hopelessly arbitrary and inadequate. Why is it ethical to use stem cells made from human embryos before August 9 but not after?

The scientific community knows that it will take years to find out whether it is possible to use stem cells to make new cells and tissues to treat diabetes, Parkinson's disease, or spinal cord injuries, and the stem cells lines arbitrarily designated by the president for use in research will not come anywhere close to being sufficient.

Private industry—which is immune to the president's ruling as long as a company accepts no federal funds for the work—knows this, too.

As the Advanced Cell Technology announcement makes clear, the race is on to find new sources of stem cells. Indeed, finding ways to create stem cells that would be completely accepted by the recipient's body—through cloning or tricking an egg to develop on its own—is where the action is in stem cell research. Unfortunately, this is not where the federal government's science is or where the government's money is.

The federal government has already fallen behind the private sector with respect to stem cell research. Private companies not only are pioneering new strategies for creating cells and tissues but also are racing one another to nail down patents and ownership of these technologies.

Without a strong federal presence in stem cell research, the American people will have to wait longer for their cures—and pay more when they finally do arrive.

And what about the problem of making embryos? Aren't cloned embryos or those created from a woman's egg the same as other embryos?

No, they are not. And this is where the president and Congress need to be very careful about what they do in response to the Advanced Cell Technology announcement.

Most Americans believe that human cloning—making a person by cloning—is going to happen soon. But the scientific research done to date does not support that belief. No one has even been able to clone a dog or a cat or even a primate from adult cells using Dolly-style techniques. Although a rhesus monkey has been created using fetal cells, most people don't have a fetus from which to extract cells to clone.

The older the source cell, the greater the risk of getting a deformed or dead embryo or fetus. So that is why adult cloning for reproduction is dangerous and may be inherently too dangerous to use. And why it may well be that despite all the efforts to make cloned human embryos, none of them—none—can grow into a person.

If cloned human embryos or those made by tricking an egg into developing cannot become people, then what is the ethical objection to creating them and using them for stem cell research?

In other words, if these embryos lack the ability or potential to become people, then they may be the best place to turn to get cells that can treat and heal.

Nothing Advanced Cell Technology has done answers the question of whether all embryos have the potential to become people. Much more research is needed.

Until it is done, the best thing Congress and the president can do is to put a moratorium on making people using human cloning, while reinstituting federal funding of stem cell work and leaving the basic research on new ways to create human embryos alone.

Korean Cloning Fraud

THE NEWS ABOUT the South Korean team involved in announcing the world's first successful generation of stem cells from cloned human embryos has gone from bad to worse to, well, awful.

Last year the South Korean team and its lead researcher, veterinarian Hwang Woo Suk, were at the top of the scientific world. In February they published a paper in the journal *Science* announcing the world's first human embryo cloning. The South Korean group, working with Gerald Schatten of the University of Pittsburgh, reported cloning thirty human embryos. The paper went on to say that stem cells had been successfully extracted from the cloned embryos that were genetically matched to specific people with various diseases.

Hwang became a huge celebrity in South Korea. The government poured money into his lab. The South Korean government was so excited that the nation had the lead in this promising area of medical research that it announced this past August that it was investing significant funds to create a world stem-cell hub under Hwang's direction that would supply stem cells from cloned embryos to researchers worldwide.

Now Hwang is in a hospital being treated for extreme stress. He has acknowledged lying about where he got the eggs used in his research, both to the public and to the journal that published his work. His key American collaborator wants nothing to do with him. One of Hwang's closest colleagues says key data reported in the *Science* paper was fraudulent. Hwang is denying these claims, but further investigations into irregularities are under way.

Predictably, opponents of stem-cell research are delighted by Hwang's disgrace. Richard Doerflinger, deputy director of the Secretariat for Pro-Life Activities, United States Conference of Catholic Bishops, burst into print to hyperventilate that "the fact is that the entire propaganda campaign for research cloning has been filled with misrepresentations, hype and outright lies."

When the issue is embryos, Doerflinger and other critics of stem-cell research are quick to forget about the merits of the science or the needs of the sick and to use the Korean scandal to impugn what their moral compass unerringly tells them must be abhorrent—seeking to take stem cells out of human embryos made from a human egg and DNA from a skin cell.

Yelling ethical fire over the Korean fiasco in a world full of people dying of incurable diseases and plagued by cruel disabilities may play in some circles, but is unlikely to fly as a guide to the real lessons to be learned from what happened in South Korea. What should be learned?

Science depends upon trust in the honesty and integrity of its practitioners, perhaps more than any other human endeavor. The peer-review process means that experts review the data and the methods that are reported but, ultimately, scientific articles are a form of testimony. Other scientists then attempt to repeat research results—so that if something is fabricated it will typically be discovered over time. But oversight committees like human experimentation committees have to trust what their investigators tell them. Journals also must rely on scientists to tell the truth to them. And researchers themselves often have to rely on the honesty of their graduate students and postdoctoral students. When trust breaks down, the very possibility of science is threatened. That is why so much time is spent these days emphasizing to young scientists the importance of integrity.

The mess in South Korea reveals the ethical problems inherent in high-pressure, high-stakes, and highly competitive science. Ever since James Watson described in his book *The Double Helix* the shenanigans that he and his collaborator Francis Crick engaged in to be the first to discover the structure of DNA it has been very clear that ambition, competitiveness, and the desire to be the first can lead the best biomedical researchers to engage in dubious, immoral, and even fraudulent behavior.

We do not know yet if Hwang and his colleagues lied about deriving cell lines from cloned embryos, but we know they lied about how they procured the eggs, and that alone is grounds for severe censure.

Honesty and integrity are partly individual traits, but they are also a reflection of institutional culture. If we create a system where there is overwhelming pressure to succeed at all costs, we should not be surprised when corners are cut. One of the keys to preventing future scandals with respect to stem-cell research will be creating a culture within institutions in which integrity is taught and nurtured and all participants are respected. Adequate mentorship by our scientific leaders will also be critical. Above all, training in ethical standards must be seen as central to the enterprise of science rather than burdensome make-work.

The real lesson of the scandal in South Korea is not that embryonic stem-cell research cannot be pursued ethically but that it can only be pursued ethically.

Media Bungled Clone Claim Coverage

DO NOT BELIEVE that chemist Brigitte Boisselier and her cloning company Clonaid, which is sponsored by the manifestly crazy cult known as the Raelians, have created a clone. I do believe that a number of negative ramifications have resulted since Boisselier appeared just after Christmas in a tacky Hollywood, Florida, motel room to announce to the world that the first cloned human had been born. And the blame for these unfortunate events must be laid squarely at the feet of the media.

As soon as I heard about the Raelians' cloning claim, I knew it was nonsense. The group has no scientific or medical experience, published no articles or reports in any peer-reviewed journals related to cloning, and produced absolutely no proof of their claim.

Cloning has barely worked in animal species—maybe one in one hundred live animals have been born per attempt—and a number of animal species have proven impossible to clone at all, including dogs and all primates. Clonaid officials' claims that the company has been successful in five of ten attempts are simply incredible on their face.

And now Clonaid is beginning to waffle on whether there will be any DNA testing allowed on the alleged clone baby, thereby making its claim completely worthless.

What's the Harm in a Little Fun?

Soon after the story broke, I began a week of appearances on television and radio and gave a host of newspaper interviews, all in an effort to try to debunk Clonaid. As I did so, I became increasingly angry about the media coverage of this nonevent.

Some might ask, "So what if media coverage was bad? What difference does it really make if these latest cloning claims are simply a cult's way of raising money and recruiting new members? The cult members make for great television with their leader's 'Starfleet Command' uniform and Boisselier's exotic appearance.

"What is the problem with giving some airtime to what at times is a vaguely amusing story about a UFO cult and its cloning fantasies?"

Unfortunately, when the media give voice to a disreputable cult in this way, great harm is done.

For starters, the cult is using the media both to raise money from vulnerable people and to recruit new followers. In addition, some antiabortion advocates, including President Bush and key Republican leaders, are using the cult's claims to advance their agenda to ban all types of cloning, regardless of whether it's for reproductive purposes or vital medical research.

To top it off, fringe scientists have been able to enhance their status by beating up on one of their own.

And, at the end of the day, the public comes away from the Raelian cloning story terrified by advances in genetics, the very science that holds the key to solving some of the biggest challenges human beings will face in this century.

Public Left Confused and Clueless

Despite twenty-four-hour media attention to the story, the American people have been left confused, scared, and clueless.

Most Americans now believe that human cloning either has been done or will be done very soon, whereas most experts believe the opposite.

One example of this kind of misleading reporting is William Saletan's online article in *Slate* on December 31. Saletan writes, "Most scientists doubt Eve is a clone, but they agree on two things. First, the various groups that have been trying to clone a human will succeed pretty soon, if they haven't already."

The public also has been told that cloning devalues life, is linked to abortion and is a tool to raise the dead as well as to manufacture armies of clones. For an example, take a look at Cal Thomas's article "Why Not

Cloning?" published by Tribune Media Services on January 1. It is sheer nonsense.

Americans were not being told that Boisselier, Clonaid's chief scientist, is a chemist with no background in medicine or biology. She has never been published in a scientific journal related to cloning, given lectures on cloning, or shown any expertise in cloning.

In addition, some of the so-called scientists used by the media— including CNN, MSNBC, Fox, and the *New York Times*—to challenge Clonaid's claims score nearly as high on the "fruitball-ometer" as Boisselier and Rael. They include the discredited Panos Zavos and the kooky Italian clone-scientist-wanna-be Severino Antinori.

Scientific Standards Ignored

The message the public receives from this indiscriminate coverage is that science has no standards. CNN aired Boisselier's December 27 "news conference" live. She produced no mother, no baby, no DNA test, no description of cloning methods, and no independent corroboration. In short, no proof—no nothing.

There is no scientist who could get this kind of coverage with this kind of rambling drivel. But the Raelians did.

As soon as Clonaid said it had no proof to present, any serious media coverage of the story should have ended.

Also, the public has barely been told that Clonaid has a history of fraud. In 2001, it managed to scam Mark Hunt, a lawyer and former West Virginia state legislator whose ten-month-old son had died of heart disease. Hunt spent $200,000 on a project backed by Clonaid to bring his baby back. Ultimately, the Food and Drug Administration shut down the program, which clearly could not have cloned anything given that the expert involved was a graduate student with no background in cloning.

Policy Ramifications

The American people are facing a real political choice about cloning. They must decide whether it should be used for conducting stem cell research or for making babies or people, or not at all. In light of the media coverage of the Raelians' cloning fantasies, the public could not possibly understand that choice.

To further confuse the issue, the media consistently allow politicians to get away with fuzzing their view on research as being opposed to reproductive cloning. When lawmakers say they are for a ban on cloning, any journalist worth his or her salt should be asking, "For research, too?" Most have not.

The media have shown themselves incapable of covering the key social and intellectual phenomena of the twenty-first century—namely, the revolution in genetics and biology. This revolution is sweeping into medicine and will soon revolutionize our understanding of human nature and behavior. It will fundamentally alter the way we make plants and animals, may cause us to rethink how we reproduce, and offers the prospect of improving or enhancing our genetic makeup.

But despite the enormous importance of these issues, most Americans now associate genetics with a man with a ponytail in a white outfit who thinks that he can live forever by downloading his memories into a cloned body.

We must learn from this fiasco.

The media simply have to do a better job in reporting on science and medicine. Public policy on cloning or other possibilities presented by our exploding knowledge of genetics cannot be based on the pronouncements of cults, kooks, and con men.

Chutzpah

NO MATTER YOUR ethnic background, you have probably heard the word *chutzpah*. The classic way to explain this Yiddish term is that chutzpah is murdering your parents and then throwing yourself on the mercy of the court asserting that you are an orphan.

Chutzpah is what is very much on display these days when it comes to those defending President Bush's position on stem cell research. They are howling in print and through the airwaves that proponents of stem cell research are not treating the moral issues involved with appropriate seriousness and complexity. Charles Krauthammer, a member of Bush's bioethics advisory council, whined in a recent issue of *Time* magazine that "the way Democrats have managed to caricature and debase the debate over embryonic stem-cell research stands in a class by itself." William Frist, Bush's doctor on call in the Senate, thunders "shame" on the critics of the president for misconstruing his position. Laura Bush has complained that "my dad died of Alzheimer's, and to hear people say that cures are right at our fingertips, it's just not right." Now, friends, Mr. Krauthammer, Senator Frist, and the lovely first lady are all guilty of chutzpah. As hard as they try, it is the defenders of the president's ban on funding for embryonic stem cell research who make a mockery of the moral issues involved. The president's policy makes no ethical sense whatsoever.

True, President Bush did not wake up one day and decide to oppose stem cell research. Indeed, as his outraged defenders note, he thought about it—for a few weeks. The president consulted with various ethics experts, albeit a majority opposed to stem cell research. On August 9, 2001, in his first major speech to the American people, he announced that he had settled on what he pronounced a "compromise" with respect to embryonic stem cell research. No more embryos could be destroyed in the name of any research involving funding by the federal government. But, drawing what can only generously be described as an arbitrary moral line on the calendar, he did allow research with any embryonic stem cells already in existence as of the day of his speech. He said there were more than sixty such cell lines around. And federal money would be available to support research on these cells.

Now, when this ban on embryonic stem cell research is correctly and accurately described as a ban, supporters of the president such as Krauthammer and Frist get exceedingly agitated. The president, they insist, compromised. "Use what was around before 2001" is not a ban. And, they self-righteously point out, this president has put in more funding, nearly $25 million, for embryonic stem cell research, than any other president.

Well, sorry, fellas, but prohibiting the expenditure of federal funds on embryonic stem cell research after 2001 is a ban. It is a ban of limited scope, but a ban it most certainly is.

Moreover, as anyone who has bothered to follow this issue knows, there are nothing like sixty stem cell lines around. There are actually twenty-three. No credible scientist believes that the number is sufficient to undertake a serious clinical research program on embryonic stem cells.

And the amount of money the president has allowed to be spent for embryonic stem cell research—$25 million—is about the size of the budget that has been allocated to study complementary and alternative medicine. The president's alleged compromise has produced funding on a par with funding for studies of St. John's wort, tai chi, and black cohosh root. Or, to put it another way, the $25 million the president has spent is about the amount of federal grant money a large department of medicine or neurology at a single major medical school might get in a year or two.

Not only is the president's "compromise" nothing of the sort, but his moral reasoning, and that of his defenders, is at best obtuse.

The president says embryo destruction is wrong but still allows research on embryos destroyed before August 2001. Huh? The president says embryo destruction is wrong but does absolutely nothing to prevent the daily destruction of embryos in infertility centers around the United States. What? The president says embryo destruction is wrong but fails to tell us whether he really believes that an embryo destined to be destroyed at a clinic but now residing in a dish is morally on a par with a child suffering from juvenile diabetes or a person who cannot walk due to a spinal cord injury. Really? And the president says embryo destruction is wrong but does not tell us what he proposes to do about American scientists going to conduct research involving embryo destruction in Korea, Britain, China, or Singapore and then publishing the results in American journals and seeking American patents. Why?

Just to drive home the point that the president insists on a ban on stem cell research, consider his position on cloning. There is another source of embryonic stem cells besides human embryos. You can create cloned embryos as sources using the techniques involved in creating Dolly the sheep to make them in dishes. His defenders say the president stands for compromise. Really? This president has done nothing but vigorously try to ban this method for getting stem cells, and while he otherwise has little time for the United Nations, he is currently devoting much energy to trying to persuade that body to ban research cloning. Some compromise.

But, wait—isn't it true, as Laura Bush points out, that proponents of embryonic stem cell research have overhyped it? Yes, that is true. But every form of scientific research in twenty-first century America gets overhyped. And if you are faced with a president who is trying to ban research that you consider vital to finding cures and therapies, you might do a bit of overhyping yourself. And if you are Laura Bush, you certainly must know that your husband's policy of banning stem cell research is the cruelest thing you can do to those with incurable diseases.

What is really going on here? The president does seem to be a person who believes that human life and human rights begin at conception, even if conception occurs in a dish. The president and his operatives know that their core base of supporters fervently oppose abortion. They also know that there is no chance that the American people would enact a law to give full legal standing to human embryos. Whatever Americans think about embryos, they do not think of them as people.

So, they made a political calculation to use stem cell research and cloning as a low-risk stalking horse to advance their antiabortion agenda and secure support among their most avid prolife constituents. And they thought that cloning gave them the chance they never had—a way to get the embryo recognized as a person in American law by banning research involving cloning. They gambled that by terrifying the public about the horrors of human cloning and by talk of embryo destruction, they could get traction in public policy with respect to abortion.

But this gamble failed. Embryonic stem cell research and cloning research have drawn huge interest from the biomedical community, the biotechnology industry, and, most important, patient advocacy groups and the tens of millions of Americans with serious diseases and ailments they represent. The president has staked out a position—banning federal funding of embryonic stem cell research as of August 2001 and of cloning for research—that could prove so unpopular that it might cost him a close election. So suddenly he and his supporters have recast his position as a compromise.

Let's get past the chutzpah defense and talk turkey. The president favors a ban on nearly all embryonic stem cell research. His talk of compromise is just that—talk.

The End of the Embryonic Stem Cell Debate

WHILE THE MEDIA apparently cannot bring themselves to say so, and some right-to-life and religious leaders may never be able to admit it, the frenzied and sometimes overheated debate about embryonic stem cell research going forward in the United States is over. This may seem odd since it was Bush, who has long opposed embryonic stem cell research, and not Kerry, a vocal proponent, who won the election. But the debate ended as a result of other votes in November—one in California, the other in New York at the United Nations.

The citizens of the state of California voted overwhelmingly for Proposition 71 on November 2. The vote was a direct rebuttal to the Bush administration's de facto ban on funding for embryonic stem cell research, including especially no funds for the cloning of stem cells for research. By a vote of 60 percent to 40 percent, Californians allotted a staggering sum of money, $3 billion, to pay for embryonic stem cell research done within the state. The state will issue bonds that will soon have the state spending $300 million a year for the next ten years on embryonic stem cell research. California in a single year will spend ten times what the federal government has spent to date on stem cell research.

Hard on the heels of the California decision to end-run the Bush ban came another bitter defeat for the president. The United Nations, despite enormous pressure from the United States, abandoned an effort to enact a treaty that would ban all forms of human cloning, including cloning done to create stem cells for research—something the California initiative explicitly allows. During an August speech at the UN, the president had called for a total ban. But, on November 19, the General Assembly, under pressure from Great Britain, South Korea, Belgium, and a number of other nations already aggressively pursuing stem cell research, gave up its attempt to secure a ban. Instead, the UN agreed to consider a nonbinding declaration regarding cloning. This will be couched in vague terms and will be enacted at the earliest next February. The UN voted against doing what the president has been unable to get the Republican Congress to do—ban the use of cloning for stem cell research.

Much has been made of the key role played by those who cared deeply about "moral values" in President Bush's reelection. But conservative values did not prevail on the issue of stem cell research. Voters in California rejected the president's position that embryonic stem cell research is immoral because it involves the sacrifice of human embryos. Instead, they took the moral view that human embryos should not be given moral standing equivalent to actual persons with all too real diseases that might be helped by embryonic stem cell research. In doing so, they now have guaranteed that embryonic stem cell research will proceed. They also gave the lie to the view that only Republicans are capable of addressing moral values.

The enactment of the California stem cell initiative, Proposition 71, is causing ripples all over American biomedicine. Some scientists who had set up partnerships with companies in Singapore, China, Korea, or Great Britain are now wondering whether California will prove a more hospitable environment for their work. Some California schools joke that they will have to stake out the airports to see which scientists or biotech CEOs are thinking about relocating. And states with a strong biotechnology sector are in a panic trying to figure out how to prevent California from reaping a scientific and economic bonanza from the Golden State's huge investment.

In Massachusetts, Minnesota, New Jersey, and Illinois, legislators are scrambling to enact legislation trying to make their states look attractive to scientists, venture capital firms, and biotech companies that want to do stem cell research. Wisconsin governor Jim Doyle has announced a $750 million initiative in stem cell research to try to protect the state's early lead in embryonic stem cell research. More states will undoubtedly follow as it becomes clear that if you are in a state that is seen as inhospitable to stem cell research, you are in a state that will not be able to quickly capture what promises to be significant economic and therapeutic returns.

Even proponents of embryonic stem cell research are somewhat confused by the turn of events at the United Nations and in California. Some of those involved with patient advocacy groups want to continue to press the president to drop his prohibition on federal funding for new cell lines

or cell lines derived from cloning. But the size of the California investment means that political interest in Washington in the subject is likely
to wane. The president can tell his supporters that he still supports a ban,
knowing all the time that the United States has now become a major
player in stem cell research. If ever there was a slick resolution to a very
tough political dilemma, the California funding bonanza is it.

So now the ball is in the court of the scientists who have fought so
hard to get more funding. Between the vote at the United Nations and
the California initiative, there is now plenty of money available for
embryonic stem cell research. The question is, Can science deliver on the
promise over the next ten years?

The battle over embryonic stem cell research involved a considerable
amount of hype, distortion, and dissembling. Conservatives used public
fears of cloning to confuse the issue over cloning for research. Proponents
of embryonic stem cell research responded to the unfair charge that they
were killing for cures with promises that diabetes, paralysis, Alzheimer's,
and Parkinsonism would become memories if only the research were
allowed to proceed.

Few in the debate seemed to know or care that hundreds of embryos
are destroyed every day at fertility clinics around the United States and
that tens of thousands more lie unwanted in frozen storage with no fate
other than to be destroyed. Both sorts of embryos could and should be
used in research.

In the end, common sense prevailed. Americans' admiration for
technology overcame their moral reservations about using embryos in
research. And the international community realized that by allowing
cloning for research, there was no reason to worry that a cloned human
being would be living down the street anytime soon. President Bush
may have held some of the high moral ground in the recent election,
but the conclusion of the embryonic stem cell debate shows that he did
not hold it all.

Part VIII

Mapping Ourselves

Ethics First,
Then Genetics

ONE OF MY physician friends likes to ask me whether there are any one-handed ethicists. He means whether there are any ethicists who aren't always saying "on the one hand, on the other hand" about every issue. I will be a one-handed ethicist and suggest some areas of controversy concerning the application of genetic knowledge where I think agreement can be reached about what is the right thing to do.

Genetics raises some tough issues, but they can be solved. Some years ago, I got a call from a neurologist at my medical school's Huntington's disease clinic. That's the terrible disease that causes neurodegeneration, leading ultimately to death. If you inherit the gene for this disease, you will get it. A test has been developed to see whether people have inherited that gene. The man on the phone said, "Art, we've got an ethics problem here. We've got a guy in the clinic that came to get tested. His dad has Huntington's disease. We tested him, and he's negative; he will not get the disease." I said, "What's the ethical problem?" He said, "The problem is, the genetic test we used shows us that the guy he thinks is his dad is not."

Genetics is different from other types of information because it can tell you information about other people. It tells you things about your biological relatives; occasionally it can tell you things about a group that is biologically related to you. Morally, this case makes it clear that people should not get information that might disenfranchise them or disempower them. My view was that we would tell the patient that he was not at risk for the disease but say nothing more. But I also said we're going to change our informed consent form to let people know genetic testing can reveal information they do not want to know or others to know.

Two ethical principles here are very important in thinking about how to use biotechnology in medicine. Genetic information should not be used to disadvantage people. And genetic information should be the subject of choice.

If you follow those two values where they lead, you will see obvious implications for genetic privacy; you will see implications for how we should control the use of genetic information in insurance, employment, and schools. There is work to be done in our legislatures to make sure that biotechnology empowers people and that they have the right to choose when it will and wont be used.

Consider another case. Years ago I got a call from a doctor in Virginia. The doctor said, "I've got a couple here that has had hereditary deafness in their family for many generations. They want me to do new test that advances in biotechnology have created to see if the fetus . . . has the marker for this form of hereditary deafness. What they tell me is if the child will not be deaf, they will abort that child. They want a baby like them."

My advice was that the point of doing genetic medicine is to battle disease and disorder. Being able to hear is not a disease or disorder even if parents might feel that way. And I might add that when parents come in to do genetic testing simply to guarantee the sex of their baby, that should not be done because gender itself is not a disease.

Another key ethical principle that this case shows is that the focus of genetic testing should be disease or disorder and its amelioration, not fulfilling people's fantasies or desires. Biotechnology must make it clear that its first priority is the war on disease, not indulging people's tastes or prejudices.

Despite the worry that Americans cannot reach ethical consensus, we do have consensus; we can reach answers. We can all agree that in order to move forward with biotechnology, we have to make sure that adequate counseling is available, that priority goes to the amelioration of disease, that people get to choose whether they want testing, and that genetic information can be kept private.

Consider a company called Framingham Genomics. Framingham Genomics had a plan. The plan was to work with the data of the Framingham Heart Study. In 1949, about one-third of the people who lived in the Massachusetts town were brought in for examinations and told that scientists wanted to do a study over generations to see the effect of lifestyle on heart disease.

People in Framingham are very proud of the study. They say that they have participated as a gift to the world to conquer heart disease. The company saw a chance to use biotechnology to advance the war on disease. It wanted to do genetic tests on everyone in the study, build a database, and sell it to pharmaceutical and biotech companies to develop new drugs.

The problem was the people were not consulted. Nor were they told how money made from their fifty years of altruism would be used to help the health of others.

Today, there is no Framingham Genomics. The lesson to the biotech industry is very clear. Ethics must come first. Companies that deal in genetic information must pay attention early on to the ethical infrastructure that will guide them. How will they protect privacy? How will they obtain consent? How will they be accountable for what they are doing?

There is no Framingham Genomics for one reason: ethics failure.

His Genes,
Our Genome

NOW THE WORLD knows what I have known for a long time: A good chunk of the DNA used to map the human genome belongs to J. Craig Venter, the scientist who led the effort at Celera Genomics (racing against the National Institutes of Health) to come up with the first rough look at our genes. Years ago, when the possibility of mapping the human genome was first discussed, Dr. Venter asked my advice about the ethics of this undertaking. I said that a crucial issue would turn out to be whose genes were selected for mapping.

From a scientific point of view, anyone's genes would be fine; the detail of the first map would be relatively crude, making the tiny genetic differences that exist among human beings irrelevant. But, as I've often found—and as the current battle over cloning human stem cells for medical research is making clear—much more than pure science is at issue when genetics is the subject.

I argued that the best thing to do in selecting people to donate DNA for the mapping project was to pick a cross section of individuals representing major ethnic and racial groups and use contributions from each of them to produce the map. I was not saying this simply to be politically correct. Starting with a broad sample of humanity could teach an important lesson: Genetically, human beings have much, much more in common than not. That is why the genome map we now have stands for all humans. Concealing the identity of the donors whose DNA was used would make it less likely that there would be attempts to engage in flawed genetic reductionism by trying to match traits and behaviors with the DNA of their donors.

Now that Craig Venter has outed himself as not only a major contributor to the mapping of the human genome but also as its major ingredient, these lessons are in jeopardy. Pundits are wondering whether Dr. Venter's love of sailing or his zest for a good intellectual fight are somehow going to show up somewhere on the map. They won't. The fact is that genes, certainly at the level of structure and organization revealed in the first crude genome map, do not by themselves make a person a brilliant scientist or a cautious bioethicist—or anything else. Our genes are capable of producing a wide range of behavior, personality, and character in each of us.

Our genes are not the defining essence of who we are; there is nothing sacred about them. They are a broad set of instructions that can produce a wide array of outcomes depending on the environment in which they are activated. Human nature is not written simply in our genome. This is why proposals to ban the copying of genes through cloning or genetic engineering—which play on our fear of the havoc that might come from changing our genes—are misguided. Any particular genome, whether it is Dr. Venter's or someone else's, can produce a variety of outcomes.

I wish my friend Craig Venter had not revealed that his DNA is a huge part of the human genome map. But now that he has, it is important to know that the ideas, hopes, and dreams that led him to map the genome and to use his own DNA to do it cannot be divined in the map he helped create. If you want to know what makes Craig Venter or other people tick, then, sure, it helps to know their genetic makeup. But unless you know even more about their parents, spouses, children, friends, schools, religious upbringing, teachers, heroes, and mentors, just having maps of their genes won't get you where you want to go. We have a lot of genes in common with one another. Our differences are not simply a product of our genes.

Let's Keep Our Genome in Perspective

DURING THE PAST few years, the public saw the announcement of the mapping of two important genomes both of the human. A crude map of the human genome was announced in 2002. A much more precise and refined map, the "HapMap," was announced late in 2005. Both of these announcements made big headlines. However, another genome has also been mapped over the same time period with almost no attention. And that tells us something about our own vanity when it comes to genes.

The other creature to have its genes precisely mapped is corn smut. Corn smut drives farmers crazy since it causes a fungal disease that wreaks havoc on this important crop. In typical human fashion, we flatter ourselves into thinking that no blueprint, no code of codes, no secret biological pattern could be of more importance, more significance, more all-out coolness than ours.

But in the short run, you would be smart to put your money, if you wanted to bet on which genome will prove more important, on corn smut.

It is doubtful that any scientific breakthrough in recent memory got the attention that the announcement that the human genome had been roughed out. Then-president Bill Clinton had the principal scientists involved, Craig Venter and Francis Collins, at his side in a Rose Garden press conference, coordinated with British prime minister Tony Blair and various British scientific luminaries.

The press engines of Celera Genomics Corporation and the National Institutes of Health, the home institutions of the two scientists, went into overdrive pumping out the news. The rhetoric got to be pretty lofty.

Clinton said it was the discovery of the century. Others drew comparisons to landing a man on the moon. Still others said it was the greatest breakthrough in medicine—ever. The HapMap did not get quite the same hype, but it garnered plenty of attention when it was completed at the end of October 2005.

Is all the PR and fanfare justified? Hmmm, I wonder.

Having available a primitive sketch of the biological blueprint responsible for turning us into a relatively hairless biped with a fairly large brain is certainly worthy of some enthusiastic praise. But the politicians and those who joined them in pushing the rhetorical envelope about mapping our genes would have been a bit closer to the mark if they had said that the announcement is important because we are humans, and, in typical human fashion, we love to trumpet facts about ourselves as crucial to the well-being of the universe.

The politicians and bureaucrats would have been more accurate if they had announced that the mapping of corn smut by the California-based genomics company Exelixis will produce more good for humanity in the short run than anything derived from the human genome map.

We have maps of our genes but no real way to translate them yet. There should be no cause for concern that anyone will be brewing up superbabies in a clinic any time soon based on what we know about ourselves.

Not so for corn smut. This fungus has a lot fewer letters than the three billion that make us up, but the scientists who cracked this code have them all. So the prospect of quickly finding ways to kill this particular pest, which does huge damage to a key food crop, are very good. It isn't a cure for cancer, but saving 10 percent of the corn crop from rot each year isn't peanuts, either.

Corn smut is way ahead in the practical application race for now.

As the details of our genome map are filled in, humanity will catch up to the mighty fungus. Many diseases will be found to have key genetic causes. New drugs will be targeted to people with different genetic make-ups so that they will have fewer nasty side effects and more benefits. Even more time will pass before we know which genes might play a role in shaping our personality and behavior, but that will happen as well.

The announcement of human gene maps leads some to ask whether we have gone too far in peering at our own instruction manual. No one seems to have been afflicted with metaphysical angst about peeping at the genome of corn smut, but then again we simply love to talk about ourselves.

It is doubtful that more than a few people will lose their faith because scientists have got the anatomy of the human DNA in hand. Faith and spirituality have always been in tension with scientific developments, whether it was Copernicus booting man out of the center of the universe, Darwin taking humanity down from the pinnacle of evolution for a less lofty setting at the end of a bushy tree of primate ancestors, or Craig Venter and Francis Collins showing that human DNA has a lot in common with that of fruit flies and, yes, corn smut.

Don't get me wrong, I understand why mapping the human genome is a wonderful achievement. But it is important to place the genetic revolution in some sort of perspective. A lot of what will turn out to be useful for us will involve understanding lowly creatures and microbes like the corn smut. And much of what will prove to be troubling about human genetics has yet to be found. The best hope for attaining the benefits that the genetic revolution has in store while minimizing the problems is to stay humble and even a bit bemused by our power, our vanity, and—as corn smut should remind us when it foils our best efforts to grow our dinner—our false arrogance about our singular importance in the grand scheme of things.

"Darwin Vindicated!"

THE MEDIA FLUBBED the headline for the biggest news event in the past fifty years of science. The reporters and TV talking heads who enthusiastically reported the breakthrough of the mapping of the human genome did understand that this was a "big" story. But, they missed the real headline. Their stories should have simply said, "Darwin vindicated!"

If you look back at the press coverage of the discovery of the human genome, you find that most reporters ballyhooed the fierce competition between scientists working for the publicly funded Human Genome Project and those employed by the privately funded Celera Genomics Corporation of Rockville, Maryland, to gain credit for the discovery. Others wondered about the financial implications of allowing human genes to be patented.

Still other headlines were meant to give us pause about whether it would be good or bad to know more about the role genes play in determining our health. Knowing more about our genes, after all, might not be so great in an era in which there is not much guarantee of medical privacy but a pretty good chance of discrimination by insurers and employers against those with "bad" genes.

There were even a couple of headlines that suggested that humanity should not be quite so arrogant since we do not have as many genes as we thought relative to other plants and animals. In fact, as it turns out, we have only twice as many genes as a fruit fly, or roughly the same number as an ear of corn, about thirty thousand.

But none of these headlines capture the most basic, most important consequence of mapping out all of our genes. The genome reveals, indisputably and beyond any serious doubt, that Darwin was right: mankind evolved over a long period of time from primitive animal ancestors.

Our genes show that scientific creationism or its close cousin, intelligent design, cannot be true. The response to all those who thump their Bible and say there is no proof, no test, and no evidence in support of evolution is "The proof is right here, in our genes."

Eric Lander of the Whitehead Institute in Cambridge, Massachusetts, has rightly said that if you look at our genome, it is clear that "evolution ... must make new genes from old parts." In other words, the genes that we have clearly are descended from similar genes in what have to be our animal ancestors.

The core recipe of humanity carries clumps of genes that show we are descended from bacteria. There is no other way to explain the jerry-rigged nature of the genes that control key aspects of our development.

No one can look at how the book of life is written and not come away fully understanding that our genetic instructions have evolved from the same programs that guided the development of earlier animals. Our genetic instructions have slowly assembled from the genetic instructions that made jellyfish, dinosaurs, wooly mammoths, and our primate ancestors.

There is, as the scientists who cracked the genome all agreed, no other possible explanation. The very same mechanisms that allow metabolism and mobility in lower animals are present in our genes, too.

The big news associated with mapping our genome is that mankind evolved. The theory of evolution is the only way to explain the arrangement of the thirty thousand genes and the three billion letters that constitute our genetic code.

The history of humanity is written in our DNA, and it shows in every section of our genes. Those who dismiss evolution as myth, who insist that evolution has no place in biology textbooks and our children's classrooms, are wrong. Our genetic arrangements are so obviously cobbled together from those possessed by our ancestors that the case for intelligent design goes out the window as soon as anyone compares us with our primate and mammalian forbears.

The message of our genes is this: Charles Darwin was right.

Ready for the Genomic Age?

IT HAS BEEN many years since Craig Venter, the former CEO and scientific majordomo of Celera Genomics, a private company based in Maryland, said that he and his colleagues had finished analyzing all the components that make up human DNA.

Nearly every human cell contains a full complement of the DNA, the software that drives our development and aging throughout each of our lives. There are about three billion letters in the roughly eighty thousand genes that make up the instructions for making a human. Celera said it has now identified and located almost all of those three billion letters.

While it will take many more years to develop a dictionary to understand what the letters and the genetic "words" they spell mean, and many more years past that to learn how to change or manipulate the words to rewrite our programming, the mapping of the genome is one of the most important things that will happen in this century.

But, sadly, the potential value of this monumental achievement may be delayed or even lost if we do not move public policy and the law forward to respond to what science has achieved.

Forgotten in all the hoopla about cracking our own DNA code is that we have done almost nothing to prepare for the advent of a flood of new genetic information about ourselves. The most immediate application of new DNA knowledge will be to correlate patterns of DNA with disease states. This has already begun with early findings of patterns of inheritance with respect to cystic fibrosis, breast cancer, prostate cancer, and Huntington's disease. These have all been linked to specific typos in our DNA.

Yet, despite years of ethical hand waving about the dangers of having DNA information without a solid public policies governing how this information can be used, no policy exists. We've had some movement regarding patents and intellectual property, with a higher standard being set for what is required to patent a gene than existed a few years ago, but, while important, much more is needed.

There is still no guarantee that anyone who wants to use genetic testing to find out about his or her risks of disease will be able to get their insurance company or HMO or federal insurance to pay for the cost of the test. In general, in the American health care system, unless something is a treatment, it is hard to get it paid for. Genetic testing could become a key part of a strategy for preventing disease and thereby lowering our health care costs, but it won't if insurance companies won't pay.

There is no federal protection should an insurance company or employer decide to use genetic testing to discriminate against those whose genes put them at high risk of getting a serious disease or a costly disability. Some states have passed laws, but we need a national policy that applies to everyone. IBM has said it will not discriminate against its employees for having bad genes, but almost no other company has followed suit.

Genetic privacy is in no better shape. No international privacy statute has been adopted to specifically protect genetic information. There is nothing to stop those who have tissue samples or biological materials stored from going out and looking at them to see what they can learn about individuals. No prohibition says the dead cannot be sampled by those curious about their genetic makeup. Companies are left to make up their own policies about consent and control of stored genetic information.

Nor are we prepared with the professional expertise we will need to translate the genetic revolution from the lab to the doctor's office. Who is it who is going to explain your genetic test results to you?

We are woefully short on trained personnel to counsel us about what genetic tests mean. Nor is there any agreed-on standardized training in law and ethics for those who will be the point persons in using genetic information to diagnose and treat you and your children.

We have not even set any standards for how marketing and advertising should work in the realm of genetics. We don't even have consensus on how accurate a genetic test needs to be before it can be sold to your doctor or directly to you. The opportunities for exploiting fear and worry are real—and it's no stretch to imagine panic as a result.

There is every reason to celebrate the triumph of humanity deciphering the component parts of its own biological programming. But, if we are going to enjoy the medical and public health benefits that this work can bring, we must get moving quickly to build ethical and legal protections that will ensure that this knowledge will be put to our collective benefit.

Unethical Policies Undermine Value of Genetic Testing

SCIENTISTS HAVE now found a gene that's responsible for the most deadly form of breast cancer: inflammatory breast cancer, or IBC. IBC accounts for just 6 percent of the breast cancer diagnoses in the United States each year, but this form of cancer is much deadlier than other types of breast tumors. Only 45 percent of women with IBC survive more than five years.

The bit of DNA responsible for this virulent form of breast cancer is called the RhoC GTPase gene. All women have this gene, but those with IBC have too many copies. This causes the body to overproduce a chemical that lets cancer cells grow rapidly and spread quickly.

With this discovery, it will soon be possible to test women at risk for breast cancer to see if they have too many copies of the IBC gene. This could lead to earlier and more effective treatment.

So far so good. But not all is as it should be when it comes to incorporating new knowledge about the role heredity plays in causing breast cancer.

In England, a law has been passed that permits insurance companies to use the results of genetic tests to refuse coverage or set higher premiums for those born with "faulty" genes that could produce premature death or costly medical bills. This may not be the best policy to follow in order to get people to take advantage of the opportunity to have genetic testing.

The Genetics and Insurance Committee, which reports to the Department of Health, gave approval, despite advice to the contrary from other panels and groups, to make Huntington's disease the first disease for which insurers can factor in genetic risks into their rates.

Letting insurance companies use genetic test results may not create a huge problem in England, where all citizens have access to the National Health Service. But should we decide to follow Britain's lead and allow our private insurance companies, managed care organizations, and employers to use genetic test results, it will create havoc.

First, many people who might consider using genetic testing to see if they are at risk of diseases like inflammatory breast cancer might have second thoughts. Fears about their loss of privacy and the possibility of losing their insurance could scare them away from tests that could save their lives.

And second, such regulations could wind up leaving not only those who are tested at risk of losing their job or insurance but also anyone who is a direct genetic relative, such as a sister or daughter. Genetic testing, unlike other forms of medical tests, reveals information not only about the person being tested but also about their relatives whether they consent to a test or not.

There are those who will argue that if people are at risk of getting expensive or fatal diseases, then it is only fair to let insurance companies know about such risks. Otherwise, what is to prevent them from loading up on health or life or disability insurance?

But the answer to how to use new genetic knowledge is not to add still more people to the ranks of those who have no health insurance or cannot change jobs for fear that they will never get insured again. If the fruits of the genetic revolution are to become available to all, we are going to have to start to reverse the long history of making insurance available according to the risks of the biological lottery and start making minimal insurance available to all.

The British have shown us exactly what not to do. The researchers who have found yet another piece in the complex puzzle of breast cancer make it imperative that we quickly reach a consensus on what is the right thing to do when biology and public policy collide. No one should forgo a genetic test for fear of losing insurance.

Who Needs
Bill Gates?

WHO NEEDS Bill Gates? No, I don't mean who needs a gazillionaire corporate titan hanging around. I am not asking why you would put up day after day with a man whose company, Microsoft, took in more than $34 billion last year by controlling nearly all the software used to run nearly every computer on the planet. No, I mean, literally, who needs him? If you could go back in time and stop the birth of the world's most famous nerd, would you?

You probably answered my question with a no. Whatever Gates's sins may be, he is the father of a computer revolution that brought much good to many people throughout the world. Add to that achievement his current generous philanthropic activities supporting some very worthy causes, such as vaccine research and a center for autism research in Seattle, and the case for having Bill with us becomes pretty persuasive.

But what if I told you that it's possible that Bill Gates has a medical condition that accounts, in part, both for his special achievements and for his "nerdiness"? Would you, if you had been his potential mom or dad fifty years ago, want a child that would go on to do great things but would have a hypernerdy personality? What if you could have known Bill's abilities and flaws before he was born? What if the decision about whether to have a child like him also carried a risk that he might be born with far more serious disabilities?

The reason I ask the question about allowing or preventing the birth of Bill Gates is that Gates is widely reported to display many personality traits that are found in a genetic condition known as Asperger's syndrome. Asperger's might be thought of as a very mild version of a more serious condition—autism, which renders many children unable to talk, be touched, communicate, or socialize.

The issue of what if anything to do about children who are severely or simply mildly autistic is of special interest for two reasons. A new patients' right group—Aspies for Freedom (the reference here is to

Asperger's)—is pushing to make June 18 Autistic Pride Day. In their view, those with autism are no more suffering from a disease than are those who are unusually short or have darker skin than others. They want autism treated as merely a difference, not a disease. And they are aghast at the thought that anyone might try to prevent the birth of a child because that child might have any degree of autism. In addition to this push for civil rights for those with autism, the chances are good that we will soon have a genetic test for detecting the risk of autism in either an embryo or a fetus.

There has been an explosion in kids diagnosed with autism in this country in the past few years. Less well known is that there is a parallel autism epidemic in other countries such as Ireland and Great Britain. Whatever the reasons for the increase in the number of cases, there can be no doubt that autism has a genetic component. Scientists and doctors have not nailed down exactly what the genetic contribution to autism is, but the fact that males are far more likely to be affected than females and that autism appears in certain ethnic groups more than others is a clear indication that there is a strong genetic component to autism.

Like many genetic diseases, there is a broad range of severity with the condition. And like some genetic diseases, say, sickle-cell trait, there can be, in the right environment, an advantage to having a mild form of a genetic disease. Asperger's is the least disabling form of autism. Research is beginning to show that it may also account for the presence of some special capabilities in areas like mathematics, computer science, and engineering. But the same genes may also create a person who is socially awkward, easily distracted, very introspective, and in many ways withdrawn and solitary. While I certainly do not know whether Bill Gates has Asperger's, his difficulties in social settings are nearly as legendary as his genius, so he might.

Gates was born on October 28, 1955. When he was born, the science of human genetics was truly in its infancy. Newborn babies were tested for a few rare genetic conditions. Fifty years later, the field of human genetics is thriving. Tests have been established for detecting Tay-Sachs disease, the gene for Huntington's disease, some forms of breast cancer, some forms of Alzheimer's disease, and hundreds of other lethal or disabling conditions. And the drive to find more tests continues unabated. Undoubtedly, the genes for autism and for Asperger's will soon be found. When they are, my question—whether you would have stopped Bill Gates from existing—takes on a very real meaning.

There are those in the autism and Asperger's community, like the newly formed and vocal Aspies for Freedom, who worry that the minute a genetic test appears, it will spell the end for a lot of future Bill Gateses. Maybe there will be fewer Thomas Jeffersons or Lewis Carrolls—remarkable thinkers who also fit the profile of Asperger's. By testing to predict a risk of autism, doctors and scientists might also eliminate the much milder and in many ways advantageous version of the condition—Asperger's.

We are going to face some very tough questions as genetic testing moves into the world of mental health. Will medicine suggest that any and every variation from absolute normalcy is pathological? How can we draw lines between disabling diseases such as severe autism and differences such as Asperger's that may give society some of its greatest achievers? Will parents have complete say over the kind of kids they want to bear? What sort of messages will doctors and genetic counselors convey when we are talking about risks, probabilities, and choices that involve not life and death but personality and sociability, genius and geekiness?

All I can tell you is that neither medicine nor the general public are at all ready to deal with the emerging genetic knowledge about autism, Asperger's, or other aspects of mental health. But the future of our society hinges on how we answer these questions.

Part IX

Reproduction

Let's Talk about Sex

THERE MAY BE a sillier strategy for dealing with sex among teenagers than promoting the choice of abstinence-only-until-marriage, but I am not quite sure what it is. Not only is such a policy contradicted by everything that medicine and science know about teenagers and sex, but it flies directly in the face of everything all Americans know about teenagers and sex.

Recent surveys show that 70 percent of all American teenagers have engaged in oral sex by the time they are eighteen. More than 70 percent of young women and 80 percent of young men approve of premarital sex, according to a just-published study in the *Review of General Psychology*. The idea that teenagers will remain celibate until they marry says much more about the people who are promoting these policies, their values, and their fantasies then it does about teenagers.

So what should we teach our kids about sex? Most Americans want young people to be taught about sex as part of their junior high school and high school education, but there is almost no agreement on what they think the content of sex education should be. It ranges from "just say no" to how to find a female's G spot. And since sex brings out our personal morality like almost no other subject, science and the facts about sexual behavior tend to get lost in a lot of finger pointing and teeth gnashing.

If you live in North Dakota, sex education is encouraged, but there is little said about what should be taught. If you live in South Carolina, state law severely restricts what can be taught. There can be no discussion of contraception except with reference to marriage, no discussion of abortion, and nothing said about homosexuality except with reference to preventing sexually transmitted diseases. Texas, ever since the days when George W. Bush was governor, has been a leader in the abstinence-only approach to sex ed.

Oregon, California, and New Jersey, on the other hand, mandate that if a school does teach about sex, it must be medically accurate, age appropriate, and respectful of the diversity of relationships, including those involving persons with disabilities.

The federal government over the course of the Bush administration has been firmly in the no-sex-until-marriage camp. Over a billion dollars has gone to support abstinence-only programs. The administration and Congress have played favorites with the bulk of your tax money, with abstinence-only money going disproportionately to Arizona, Florida, Georgia, and Texas. Vermont got the least. Maybe the kids in Vermont cannot hear admonitions to remain chaste amid the sound of falling snow?

I am completely against abstinence-only sex education programs for three reasons. There is no evidence at all that they work. Common sense says that they have no chance of working. And it is not clear that ethically they send the right message to young people.

Eleven states have tried to evaluate their abstinence-only programs. The results have been dismal. In Kansas, the evaluators stated that "no changes [were] noted in participants' actual or intended behavior." Evaluators of the Texas program found the same thing—no change in the number of students pledging to remain celibate until marriage. In fact, more students reported having had sex after taking an abstinence-only sex ed course than they did before taking it! There is no evidence at all that telling kids not to fool around has any more impact at school than it does when parents say the same thing at home.

Which leads us to the world of sex and common sense. There are kids who are not going to have sex in junior high or high school. There are, according to what social scientists know about teenagers, not a lot of such kids, but there are some. There are also some teenagers who are going to engage in homosexual acts and other nonstandard forms of sexual contact. There are not a lot of those out there, according to what social scientists know about teenagers, but there are some. An even smaller number of kids will, tragically, be forced to have sex by parents, relatives, or rapists.

The fact is that a teenager has a pretty good chance of getting involved in sex before graduating high school and a small chance of being involved in something other than consensual male-female sexual intercourse. So in addition to there being no evidence that abstinence-only sex ed works, there is no reason to believe that this form of sex education is even on the same planet as those it is intended to reach.

Which leads me to my last and most controversial claim: what message is sent when abstinence until marriage is the only acceptable way to deal with sex? When I went to student–parent meetings at my son's high school, parents of girls were frantic that the school reinforce the message that sexual intercourse was a bad choice. Parents of boys always seemed to me to be supportive but not nearly as frantic that this message be taught. But all that seemed to change when these same kids went away to college or went off to get a job. A lot of these very same parents stopped preaching that sex before marriage was wrong. A fair number of them would whisper that sex before marriage might be a good idea, especially if the sex was with someone their son or daughter was thinking about marrying. Many of them had lived with someone before marrying, and all of them who had done so had had sex before marrying.

I think the message that sex must wait until marriage is not the right message to send to a young person. First, the people sending the message almost never lived up to it in their own lives, and nothing turns a kid off like hypocrisy. Second, kids themselves don't believe it. And lastly, it is not beyond an ethical argument that marriage must always precede sex no matter what age you are. It makes more sense to talk about maturity, love, voluntariness, and mutual respect than to send an absolute message that sex is unacceptable outside marriage that gets nullified the day a person graduates from high school. Science and common sense, not wishful thinking and hypocrisy, should guide what we teach about sex.

Model Eggs

A FASHION PHOTOGRAPHER by the name of Ron Harris is offering the eggs of eight models for sale. The auction of the gametes of these beauties began on the internet, and bids started at $15,000. The shock waves from this pathetic idea have already started to reverberate in conversations around the water cooler, over the Web, and on the late-night talk shows. The idea, most people say, is disgusting. Even the American Society of Reproductive Medicine, which has never encountered a business transaction in the area of reproduction it did not like, calls the auction "unethical" and "distasteful."

If you do not like what Ron and his models are up to, you have only yourself to blame. The lack of any regulations governing what can be done with sperm, eggs, and reproductive organs makes schemes like Ron Harris's inevitable.

Harris's idea is pathetic for three reasons. First, beauty is not heritable. If it was, then beautiful kids would have beautiful parents. Beautiful people would have beautiful siblings since they share a lot of genes. More often than not, they don't. That is because appearance is the function of many genes, and, when sperm and egg mix, the combination that is produced may or may not closely resemble the persons whose sperm and egg were used. Beauty is a matter of millimeters, and it does not take much genetic reshuffling to move a pretty face to an ordinary one.

Second, beauty is sometimes a reflection of what is unusual or different. But what is unusual may sometimes be a reflection of a genetic weakness or problem. An extremely thin body may be the result of a disposition to anorexia. An unusual set of facial features may represent a bone growth anomaly. Assumedly, health ought to come before beauty, and anyone who would risk the health of a child in the vain pursuit of beauty ought pay a fine, not be paid a bounty.

Third, by putting pictures on the Web of the eight models, these women can throw their anonymity out the window. In twenty or twenty-five years, they can expect a visit from any child created from their eggs. Perhaps they do not care, but I suspect that at least some of them have not realized that the Web is not a very private place to sell your reproductive materials.

So selling beauty through genetic selection is not a good idea. Still, the idea is a natural consequence of other efforts promoting "genius" sperm or allowing infertile couples to advertise for eggs from smart women at elite colleges.

The latter practice is so common that one could probably ditch college rankings of the sort that *U.S. News & World Report* provides each year and replace them with a price list of which school's women are able to charge the most for their eggs. The more elite the perception of the school, the higher the price infertile couples are willing to pay. By this measure, Princeton, where a woman allegedly sold her eggs for $50,000, should top the list of "Best American Colleges."

As I said, we have only ourselves to blame for turning baby making into a business. By failing to ban the sale of sperm, eggs, and rent-a-womb schemes, we have allowed baby buying and selling to become a reality. It is not too late to change this state of affairs, but time is running out. If we do not regulate the open commerce in human reproductive materials soon, then other entrepreneurs with bigger schemes and more marketing power then Ron Harris will soon be up on the Internet.

Soldier's Sperm Offers Biological Insurance Policy

WHEN SOLDIERS go off to battle, they're usually told by their superiors to get their legal and financial affairs in order. No one likes to think about the possibility of not coming home, but that it the brutal reality of war. Now there's a new twist in soldiers' preparations for the unthinkable: some of the men who go to Iraq are leaving deposits at sperm banks before heading overseas.

There are two main reasons why soldiers are doing this. One is obvious: if they are killed, then their wife or girlfriend will still be able to have their baby.

The other reason is fear of injury or illness. Some military personnel who served in the Gulf War believe they were made sterile by exposure to insecticides, nerve gas, or other toxic substances. While there is no solid medical evidence to support this theory, some soldiers heading off to another war in the Mideast are choosing to play it safe and leave their sperm behind.

What about Female Troops?

Should they die or become sterile, female soldiers do not have the same option as men to preserve their reproductive abilities. While sperm are relatively easy to freeze and preserve, the same is not true for eggs since the freezing process easily damages them.

The only real option female soldiers have is to take hormones, create many eggs, and have them surgically harvested and then quickly fertilized. They could then store one or more embryos, which unlike eggs, are easy to freeze.

The prohibitive cost of this procedure—tens of thousands of dollars as opposed to a few hundred dollars for sperm banking—means that taking out a biological life insurance policy is an option that is really only practical for men right now.

Ethical Issues Unanswered

Although the practice of freezing one's sperm before going to war has started to catch on, a host of ethical issues have not been adequately addressed:

- Should every male soldier be given this option? The Department of Defense routinely advises soldiers facing possible combat about the importance of having a will, but it does not routinely talk about sperm banking.
- Should Uncle Sam offer to pay for a soldier to store his sperm?
- If sperm is banked, then what rules should govern the disposition of sperm if a soldier does not come back? Should only wives have access to the sample, or should girlfriends, fiancées, or even family members who might want to hire a surrogate mother?
- Given that the man who made the deposit is now dead, should someone seeking to become pregnant from the sperm have to wait and think about that choice for days, weeks, or even months?
- How many children could legally be created from a deceased man's stored sperm?
- And how long should a deposit be stored for the use—one year, ten years, fifty years?

These are all tough questions for which there are no rules, laws or legislation, but the issues need to be addressed.

Technology now makes it possible to cheat death in ways never imagined by soldiers in previous centuries. We owe it to our troops to make sure that, should they choose to use sperm banking as an option, they will know exactly what will—and will not—happen if a tragedy forces their loved ones to make the ultimate bank withdrawal.

Taking Reproductive Responsibility

AS A SUFFERER of postpartum depression, Andrea Yates may not have been responsible for her behavior when she allegedly drowned her five children in Houston. But she and her husband certainly were not acting responsibly when they had another child despite her earlier bout of the disorder that reportedly led to a suicide attempt. And anyone whom this couple dealt with—family, friends, doctor, or clergy—who did not say plainly that they should not have another child did not do the right thing.

When this tragedy occurred, the local talk radio show here in Philadelphia was abuzz with comments about Yates.

The callers I was listening to did not doubt the diagnosis of postpartum depression. Unlike other claims about mental illness, this one did not seem to trigger the sort of sneering skepticism that often greets the explanation of a crime that appears to result from a mental disorder. Schizophrenics, addicts, alcoholics, and even the mentally retarded facing the death sentence often receive less sympathy from the American public than has Yates.

But during that radio show, what was interesting was the tendency of the callers to want to broaden the scope of the blame beyond Yates and her mental illness. Some callers blamed her husband, whom they said should never have left her alone at home with their children. Others blamed her doctor for prescribing her pills that might have had unexpected side effects. Still others wanted to blame the pharmaceutical companies that make the medications she reportedly was taking, saying they ought to be held accountable for the killings.

A quick review of online chat groups and call-in television and radio shows around the country revealed much the same commentary.

It is amazing that Americans seem to believe that when a person commits a crime, there must be a whole slew of other individuals and groups who are also responsible. Perhaps we have all been watching too many lawyers whose instinct is to look for the deep pocket whenever there is wrongdoing. Perhaps, despite our avowed love of self-determination, when push comes to shove, we want to find some other reason for brutality than a single individual's actions.

It may be that Yates's depression was so severe that it led to a kind of psychotic episode that could cause her to commit such a ghastly crime.

If Yates killed her kids, the real culprits are sex and America's inability to talk about sexual responsibility. Sure, we babble on about teaching our kids about abstinence and the importance of those who are sexually active to follow safe sexual practices. But the harsh truth is that we really believe that an individual's sexual and reproductive activity is no one else's moral business.

That belief is fine when it comes to public policy and legislating sexual conduct. It will not do as a matter of ethics.

You have a legal right to do what you want to do in the privacy of your own bedroom. That is sound public policy. However, if what you do in that bedroom is going to create a situation where children are going to be put in harm's way—because the parents are drug addicts, old and likely to render the child an orphan, too young to adequately care for the child, or likely to pass on a lethal illness, for instance—then each of us has an ethical duty to say that there are things you ought not be doing in that bedroom.

It is our collective failure to deal with the ethics of sex that leaves us scrambling to find someone or something to blame when irresponsible sexual behavior leads to a tragedy.

Test Tube Babies versus Clones

TWENTY-FIVE YEARS ago, one of the most revolutionary events in the history of humankind took place: a little girl named Louise Brown was born. While the birth of a baby was not unusual, this child was different. She was the first person ever created outside a woman's body. Looking back, it's hard to remember just how controversial her conception and birth were.

The world's first "test tube baby" arrived amid a storm of protest. Many in the then-emerging field of bioethics, such as Leon Kass, the former chair of President Bush's Council on Bioethics, argued that creating people by means of in vitro fertilization (IVF)—mixing sperm and egg in a glass dish (test tubes were never actually used)—was morally wrong.

Kass argued that the procedure might prove unsafe, that any technology that separated sex and the creation of life was morally suspect, that making babies outside bodies was unnatural, and, even worse, that the process treated people like objects or things.

Kass had plenty of company. The pope, various other theologians, newspaper columnists, and many doctors were also wary of test tube baby technology.

But the power of those opposed to the procedure completely vanished when Brown was born. She was a happy, healthy infant, and her parents were thrilled. The doctors who helped to create her, Patrick Steptoe and Robert Edwards, could not have been more pleased. She seemed as natural a baby as had ever entered the world.

On the day Brown was born, the naysayers and worrywarts lost their audience. IVF took off and never looked back. Tens of thousands of babies have now been born using the same method that was used to make Brown. The procedure has become so common that hardly an eyebrow is raised if you tell someone that you or your children were born with the help of reproductive technology.

Some proponents of human cloning look at Brown's birth with envy. They think the same shift from condemnation to acceptance will happen if a clone baby can be created. They argue that a clone baby will silence the critics (including, yet again, Kass). They believe all the worries about making babies safely and concerns over the loss of individuality or the unnaturalness of the technique will disappear overnight once a healthy clone baby appears on TV.

But there are two very big differences between the decision to try to make the first test tube baby and the ethics of trying to clone the first human being.

When Brown was born, scientists already had many years of experience using IVF to make a variety of animals. They had encountered few problems. But the same can hardly be said about cloning.

Nearly 90 percent of clones fail to develop into live-born animals. And among those that are born into the world alive, a huge number have serious or fatal medical problems. While Brown was part of an experiment, much more was known about the technology of IVF than is known today about cloning. And what is known about cloning is enough to make it clear that it would be the height of irresponsibility to try cloning in humans until the animal outcomes dramatically improve.

Another reason that moral criticism of IVF evaporated so quickly is that Brown and her parents did not permit her birth to become a media circus. To this day, very few people have ever seen a picture of Brown as an adult because she and her parents knew that insatiable public curiosity could turn her life into a living hell.

But those who want to clone humans are in love with publicity. They are either promoting themselves or using cloning to raise funds for their company or cult. This is the exact kind of behavior that can make worries about the creation of unnatural freaks come true in a hurry.

The history of IVF has not been without its problems. Infertility treatment has become a big business, and there are almost no rules governing who can use technology to make a baby. In particular, there are no regulations regarding the parents' mental stability or age. And, as a result of the technology, there are far too many embryos frozen in liquid nitrogen that no one will ever use to try to make a baby.

But, overall, the creation of Brown and the development of IVF was a moral success. It has brought much happiness to many. And it is one of the most prolife technologies ever created.

When thinking about why Brown's legacy has been so positive, it is important to keep in mind what it was about IVF that silenced its critics. Brown's birth was the culmination of years of solid research in animals and was carried out with a commitment on the part of her parents and doctors to put her interests first.

The Problem with "Embryo Adoption"

ONE OF THE strangest outcomes of the ongoing debate about embryonic stem cell research is the government's use of taxpayer money to support a little-known private organization called Snowflakes. Devoted to encouraging couples to "adopt" human embryos, Snowflakes has received more than $1 million from the Bush administration and Congress.

While helping people have babies is ethically commendable, there is something very strange about extending the use of the term *adoption* to embryos. Children get adopted, but . . . embryos?

And it is even stranger that the federal government is buying into this way of thinking.

So where do all these embryos that supposedly need adopting come from in the first place?

When couples seek treatment for infertility, they often wind up using in vitro fertilization, or IVF. This is a procedure in which embryos are created outside the body in a laboratory dish and are then implanted back into the woman's body where, ideally, they grow to full term.

It works like this: The woman takes fertility drugs that cause her to produce far more eggs than the one she normally would release during her monthly cycle. These eggs are then surgically removed from her ovaries and fertilized in a dish with either her husband's or a donor's sperm.

Often many embryos are created through this process. But since multiple pregnancies—quadruplets, quintuplets, septuplets, and the like—produce premature and often unhealthy babies, doctors will only put two or three embryos back into the woman's body to try to help her become pregnant.

The clinic chooses to implant the embryos that look the healthiest and asks the couple whether they want to freeze the rest. The couple also has the option of having the remaining embryos destroyed, donated to other couples, or donated for embryonic stem cell research.

"Pre-born Children Waiting"

This is where Snowflakes saw a need—and a chance to score some moral points in the debate over stem cell research.

Snowflakes is run by the Nightlight Christian Adoption agency in Fullerton, California. The group has no medical background. They simply believe that every embryo is a baby from the minute it exists in a laboratory dish.

The Snowflakes program deliberately uses the language of adoption to make that point clear. They created a service that matches couples who have leftover embryos with other infertile couples trying to have babies. To quote from their website, "By some estimates, there are over 100,000 frozen embryos in cryo-banks throughout the United States. Pre-born children waiting—waiting."

Actually Snowflakes' estimate of one hundred thousand embryos is probably very low. Most experts think there are as many as four hundred thousand embryos frozen in storage in the United States. As of just over a year ago, the Snowflakes program had received about 750 of them and had matched seventy donor couples with forty-eight other couples seeking to have children. Sixteen babies had been born.

What's the Big Deal?

So what's the big deal about a religious group that believes all embryos are children and is trying to find them "adoptive" parents among infertile couples using IVF? Well, actually, there is a lot that is wrong.

It's great that sixteen babies were born last year through the Snowflakes program. That makes is seem as if sixteen couples had children who might otherwise have not. But that is not really the case. Nearly all infertility clinics offer couples the option of donating their leftover embryos to other couples. All that Snowflakes has done is brought the rhetoric of adoption into the process.

You might also get the impression that Snowflakes is creating an opportunity for infertile couples to access the one hundred thousand to four hundred thousand frozen embryos out there. But that is not really the case, either. If you are infertile and are trying to have a baby, your best bet is not to use a frozen embryo made by a couple who had themselves been going through infertility treatment and whose embryos were not used because they did not look healthy enough.

Despite Snowflakes's rhetoric, most frozen embryos are not healthy enough to ever become babies. The chance they will grow to full term is about one in ten for those frozen less than five years and even less for those that have been frozen longer. This is why so few couples have taken Snowflakes up on its idea of "adopting" frozen embryos.

Moreover, using term like *adoption* encourages people to believe that frozen embryos are the equivalent of children. But they are not the same. In fact, infertile couples who want children can frequently make embryos but they do not become fetuses or babies.

The older a woman gets, the less likely her embryos are to become babies. For women over forty-five, the chance of her embryo becoming a baby is almost zero. The inability to make embryos that become babies is why couples turn to donor eggs or donor sperm. Almost no one who is going to spend $10,000 per try to use IVF is going to want to try it with another infertile couple's frozen embryo whose chances of properly developing grow less with every year it is frozen.

A Government Sham

The Bush administration and Congress know all these facts but have nevertheless poured more than $1 million of taxpayer money into the Snowflakes program and others aimed at facilitating "embryo adoption."

This is a nice way to score points with those who advocate the view that embryos are actual babies and should not be used for research purposes. But it is not the best way to help couples who want to have actual babies.

One million dollars would be far better spent matching fertile couples willing to make embryos with infertile couples, rather than trying to get them to use unhealthy frozen ones.

One million dollars could also help defray the staggering costs of IVF, which only middle- and upper-class couples can currently afford.

When the money is spent on programs like Snowflakes, the only explanation is ideology, not medicine.

Are You Ever Too Old
to Have a Baby?

CAN SOMEONE be too old to be a parent? A number of cases last year of single women in their midfifties having children raised this question. The question looks pretty straightforward, but it isn't. In order to answer it, there must be agreement both on what values would make a potential parent "too old" and on who would enforce a rule that used age to limit access to reproductive technologies.

So, how old is too old? Was Larry King of CNN too old when he had a child with his seventh wife at age seventy? Cheryl Tiegs, who had twins at fifty-two? Geena Davis, twins at forty-eight? Tony Randall, who had a child at seventy-seven and died a few years later? James Doohan (Scotty from the original *Star Trek*), who had children at eighty? Donald Trump, fifty-eight, now making menacing reproductive noises in the context of his widely publicized third marriage? It is very hard to draw an absolute line and say what age is too old, although the idea of Donald Trump mating makes the project a particularly compelling one.

Assistance in knowing how old is too old has now been provided by a Romanian woman, Adriana Iliescu. The sixty-six-year-old unmarried writer gave birth by an emergency Cesarean section to a baby girl on January 17 at the Giulesti Maternity Hospital in Bucharest. She is now the oldest woman ever to give birth. Her doctor, when asked why he would use reproductive technology to permit a sixty-six-year-old woman to become pregnant, said that is what she wanted to do and that he was impressed with her faith in God and with her "determination." Giving birth may be what Adriana Iliescu wanted, and she may well be a very determined person of great piety, but the doctor did something highly unethical when he helped her become pregnant. Adriana Iliescu is too old to be having a baby.

Iliescu needed a doctor's help because she no longer can make eggs. Since she is single, donor eggs and sperm were used to make embryos. One of these became her daughter Eliza Maria. But, while she did give birth, all the doctor did was allow her to become pregnant, not to have her own genetic child. Why was this so wrong? It is wrong because there was a terrible price to pay for using technology to make a sixty-six-year-old woman pregnant.

Any woman over the age of forty is a high-risk pregnancy. Medical risks rise rapidly for both moms over forty and for their babies. They became terribly real in the case of Iliescu's pregnancy.

The child she delivered was born premature —a low-birthweight baby. This poses real health problems for the baby, which are made even more troubling by the fact that Romanian neonatal units are not the equal of those in other, wealthier nations. The mom had to undergo an emergency C-section, not easy for a sixty-six-year-old, who now must take on mothering duties by herself with a baby who may well have significant medical problems.

Not as widely reported in all the hullabaloo about the "world's oldest mom" is that Iliescu lost one fetus early in her pregnancy and gave birth to a stillborn baby when Elsa Maria was born. Adriana Iliescu's pregnancy came with two deaths, one premature baby whose fate remains unknown, and one life-threatening emergency C-section—a morally unacceptable price.

But that is not the end of the challenges that a pregnancy in a sixty-six-year-old woman brings. Consider that when her daughter enters high school, her mother will be eighty. Eighty! That's the end of the argument. If you are sixty-six and single, man or woman, you should not be having a baby.

My proposal is that anyone over sixty-five who is single should not use reproductive technology to have a child. If you have a partner, then your total ages should not be more than 130. And if you are female and at or near fifty-five years of age and going to carry a pregnancy, then you can use reproductive technology only if you can pass a tough physical examination. Sixty-six—forget it.

Who will enforce this age limit? Should we have a law, or can we leave it up to doctors at individual infertility clinics?

As it happens, I coauthored a survey of American reproductive technology programs led by Andrea Gurmankin of Harvard Medical School, which was just published in the journal *Fertility and Sterility*. The survey asked a number of hypothetical questions of clinic directors to try to figure out what values doctors use to decide who can and cannot use technology to become a parent in the United States.

One question we asked was whether clinics would turn away couples where both parents were forty-three years old. Most would not, but surprisingly 18 percent said they would. Twenty percent said they would not accept a woman who was single, and fifty-three percent said they would not deal with single men.

Age was not the only disqualifier. Three percent of programs said they would not accept a couple who were both blind from an accident. Seventeen percent said no to women who state they are lesbians. Thirty-eight percent said they would not take as patients a couple who were on welfare and using Social Security checks to pay for infertility treatment.

So there is quite a lot of screening going on at clinics already. Some programs don't care much about age or whether you are single or not. Others do.

If there is an age at which it makes sense to say someone is too old to use reproductive technology, and I think there is, then who should enforce this policy? If you leave enforcement to individual clinics, they may, but their current practices, which are all over the map, make it seem unlikely. Still, the medical profession or state legislators should act. The race to create the world's oldest mom should be declared over.

Part X

The State of Science in the United States

Hullabaloo over MMR Risk Misses the Point

EVERY FEW MONTHS, I get a call from a very distraught parent. They want to know whether or not they should have their child immunized against measles, mumps, and rubella (MMR). They have heard or read on the Web that there is a link between the MMR shot and autism, a terrible brain disorder that, in its most severe form, makes it next to impossible to communicate with your child.

Now, I am not a pediatrician or even a practicing physician, so I tell these parents to talk to their doctors. But I must confess that I do tell them that I had my son vaccinated and would not hesitate to do so again.

The parents are right to be concerned about autism. The disease is on the rise in the United States. In the past ten years, the rate of kids diagnosed with autism has gone up about 15 percent every year—an almost unheard-of brain disease epidemic.

But parents panicked about autism are wrong to be concerned about the MMR vaccine as the cause of this terrible epidemic. Whatever is causing the explosion in autism, getting vaccinated against measles, mumps, and rubella is not it.

Why would anyone blame the vaccination for the autism epidemic? Back in 1998, a study appeared in the British journal *The Lancet* that found that in twelve children with autism, eight of them seemed to have shown signs of the disease after getting a measles, mumps, and rubella shot. That report set off a scare that led to many parents in the UK and the United States deciding not to get their infants vaccinated. Even the current prime minister, Tony Blair, who had called on parents to keep vaccinating their kids, had to be shamed by the British tabloid press into admitting that he had had his own child vaccinated. What is certain is that the fear this paper inspired led to outbreaks of measles, mumps, and whooping cough in the UK and the United States that could have been prevented by vaccination.

Now the authors of the *Lancet* paper have retracted what they wrote six years ago. Last month, ten of the thirteen authors of the original paper said that the data in their original paper did not support the conclusion that the vaccine was to blame.

Getting the retraction was not easy. It took a chorus of other studies arguing that the link was not there, including a 2002 comprehensive British study and a thorough report shortly thereafter by the U.S. National Academy of Science's Institute of Medicine to get the authors to admit that there paper was fundamentally wrong to connect autism with getting vaccinated.

Given all the harm and worry that the original paper spawned, why did it make it into print in the first place? Well, partly because, even though it only involved a handful of kids, it seemed to find an association that might be important, and the media took this little study and turned it into many a headline. And perhaps in part the authors were led to see more than the data implied because the main author was getting money from lawyers in the UK to see whether there was any basis for lawsuits by parents of kids with autism against vaccine manufacturers— money that he did not disclose to the editors of *The Lancet*.

The moral of this story is very clear. Everyone—the media, primary care physicians, journal editors, and parents—needs to be very careful about how early reports about medical risk are handled. It should take more than one small study to get us to stop drinking coffee, holding a cell phone to our head, eating French fries, or avoiding getting our kids effective vaccinations.

It is hard to prove what really works in medicine. It should not be easy to throw those things that really do work, like MMR vaccination, away.

If Science Becomes Politicized, Where Do We Go for Truth?

ARE SCIENCE AND technology immune from politics as usual in Washington? Hardly. During the last year, the Bush administration has been cleaning out the previous administration's panels responsible for providing expert advice to the Department of Health and Human Services (HHS) on a variety of scientific, technical, and bioethical questions. It's housecleaning with a vengeance—and I think the public deserves better than this in the relationship between science and government.

More than two hundred committees, panels, and expert bodies advise the president directly about science issues, either directly or indirectly by reporting to HHS. The Bush administration unexpectedly euthanized a committee advising the Food and Drug Administration (FDA) on how to regulate genetic testing, closed down another blue-ribbon body supplying advice on how to toughen protections for human subjects in medical research, and began revamping the membership of a committee looking at the effect of various pesticides and herbicides on human health.

The president replaced a Clinton-appointed bioethics panel with scientists and ethicists of his own choosing and overhauled a key advisory committee on blood safety and availability.

The new bioethics panel is, unlike its predecessor, decidedly more conservative and prolife. The new blood committee is, unlike its predecessor, decidedly docile when it comes to challenging industry, government regulators, and powerful blood collection agencies such as the Red Cross.

As the ex-chair of the old blood committee, I confess to the remote possibility of some tiny bias about my evaluation of the brilliance of the previous committee! But I don't think I am way off the mark, given the disquiet of patient advocacy groups about the membership changes.

Three common reactions to this sort of political housecleaning are indifference, sour grapes, and surprise. None is adequate.

Indifference is probably the most common reaction. No surprise: new administrators can the previous administration's people. Every administration wants its own folks to advise them on economics, agriculture, defense, and foreign affairs. Why should it be any different when it comes to science?

Well, it should be. This isn't the same as appointing someone from your team as ambassador to Bermuda. Science and technology today are far, far too important and, as the current concerns about biowarfare and chemical weapons make clear, too dangerous to be treated as two more arenas for political payback.

True zealots respond, Our guys are in; your guys are out—and if you don't like it, lump it. The old lunkheads are leaving, and our guys are endowed with common sense and more finely tuned moral purpose.

But this view of how politics and science should mix is wrong. The nation's interest is not well served when politicians use scientific and technical appointments to satisfy their most rabid supporters or to pay back major contributors. Policy, no matter how ideological, needs to rest on sound opinion and valid knowledge. The best test of knowledge is not a candidate's popularity with the antiabortion lobby or the anti–genetically modified foods lobby. The best test is not whether a candidate gave you votes or money. The best test is the quality and reproducibility of publications in peer-reviewed journals.

Politics may not have to follow rules, but we're all equally bound by the laws of physics. That being the case, the only relevant standard for advising about science and technology is who are the best experts. By all the important measures, many in the new group do not match up as scientists to the old group.

Those who think that science, simply because it is science, can stand outside the usual Washington political fandango are at best naive.

What the public needs is the creation of panels, bodies, and committees that do not have to be accountable to whichever party is in power. The private sector, foundations, and not-for-profit groups ought to take heed from the latest round of science purges and make the creation of independent, freestanding advisory panels and bodies a top priority. The only way the best science is going to get heard in policy circles is if there are groups and bodies that can speak freely and frankly without having to depend for their existence on the political whim of the ruling party.

Is Biomedical Research Too Dangerous to Pursue?

A **PARTICULAR ETHICAL** stance, utilitarianism, has guided science policy in the United States for more than fifty years. The moral argument that investing in science and technology extends life and improves the quality of life, despite exacting a toll in harms and risks, has long dominated debate in American science policy. This ethical framework is now under sustained attack from many quarters. Not because there are doubts about the benefits scientific knowledge can bring but because of the belief that there are important values that will of necessity be compromised if biomedicine continues in its current direction. Some making this argument have the ears of those at the highest levels of government.

During the past few years, a rash of writings has appeared that place biotechnology under a critical moral lens and reject the view that benefit alone should be the sole determinant of whether to proceed with the biotechnology revolution. As the bioethicist Daniel Callahan argues, those concerned about where biotechnology may take us do not and should not accept the framing of the debate in utilitarian talk of "progress," "cures," and "a better life." Worries about the loss of our humanity, the value of life, and the meaning of human experience are at the core of their fears.

Perhaps the most outspoken and influential critic of the utilitarian justification for research is the former chair of President Bush's Council on Bioethics, Leon R. Kass. Kass has long been concerned about the ways in which biotechnology undermines or shifts our understanding of the nature of family, marriage, sexual relations, aging, and parenting. Recent developments in cloning, stem cells, genetic testing, pharmacology, anti-aging research, and the neurosciences have only intensified his concerns.

While crude utilitarianism is surely an inadequate moral foundation for biomedical research, the case offered by those worrying about where biomedical research is headed is far from persuasive. The key moral worries of Kass and those who echo his thinking fall into three

165

areas. First, biomedical research cannot continue on its present course without significantly altering human nature. Second, if, in the name of more cures, longer life, and improved quality of life, we continue on our present biomedical research course, we will commodify and objectify human life. Third, too much biomedical tinkering will produce a loss of authenticity and meaning in human experience. We may some day feel better because of powerful drugs or immersion in a world of computer-generated stimuli, but we will be far less healthy because our sense of well-being has become programmed, artificial, and inauthentic.

Must progress in biomedicine distort who we are? Has a case really been made sufficient to slow or stop the biomedical research enterprise? I believe not.

Biomedical knowledge may lead us to attempt to tinker with the genes, neurons, or physical bodies that we understand today as defining human nature. But human nature has, especially with the rise of civilization, changed drastically in response to technology. Even the most basic ideas about who we are—how we see the world; how we walk, run, and move; whom we interact with, befriend, and love; what thrills us and threatens us—are all the result of complex interactions among the world, technology, and our bodies. Nor is there any reason to glorify a particular phase in the evolution of human nature and declare it sacrosanct. Human nature is not static; it lacks any recognized "essence" and certainly has elements that have proven maladaptive in the past.

Similarly, while we may imperil the value of humanity by seeing ourselves as portable sources of marketable organs or beings who must use technology in order in ensure optimal reproductive success, choosing such objectified or commodified visions is not an inevitable result of biomedical progress. Social and political choices, not scientific advances, will determine how our dignity and autonomy are to be squared with the prospect of birthing ourselves in artificial wombs or enhancing our mental capacities through genetic or neuronal engineering.

Self-esteem need not be a victim of progress. Those born as a result of the application of forceps, neonatal intensive care units, in vitro fertilization, or preimplantation genetic diagnosis do not appear to suffer from undue angst about having been artificially "manufactured." Nor should they. There is not anything obviously more dignified about being "made" in the backseat of a car than there is having been conceived from a pipette and a sperm donor. If commodification and objectification are the concerns that are to lead to the closing of the public purse for biomedicine, more substantive worries ought to be adduced.

Some of the most passionate reasons that critics provide as to why the march of biomedical research makes them nervous are that if biotechnology keeps moving along, it will come at a terrible price: the loss of authentic happiness, the loss of what makes life meaningful—struggle, suffering frailty, finitude, and death. But, again, the concerns do not seem to square with what we have already experienced in the wake of biomedical progress. Do those who use glasses, insulin injections, wheelchairs, inhalers, oxygen tanks, hearing aids, or prosthetic limbs feel inauthentic or overcome by a loss of meaning in their lives? If I use a calculator, a computer, or the Internet to solve a problem, do I feel that I have been cheated out of a more authentic experience enjoyed by my grandparents, who used pencil-and-paper calculation, visited a library, or mastered the multiplication table? There is little evidence for the dour view that we can be happy only when we have earned our happiness.

When the stakes are enormous—continued premature death, disability, chronic suffering—then much is required of those who would challenge the wisdom of the aggressive pursuit of biomedical knowledge that is the only hope of solving these terrible problems. Kass and other recent critics have not come close to meeting that challenge.

Misusing the Nazi Analogy

SIXTY YEARS AGO, Allied forces brought an end to Adolf Hitler's dream that Germany would rule Europe and dominate the world. The death of Nazi Germany gave birth to a charge that still haunts the scientific community—what might be called "the Nazi analogy." In ethical or policy disputes about science and medicine, no argument can bring debate to a more screeching halt then the invocation of the Nazi comparison.

Whether the subject is stem cell research, end-of-life care, the conduct of clinical trials in poor nations, abortion, embryo research, animal experimentation, genetic testing, or human experimentation involving vulnerable populations, references to Nazi policies or practices tumble forth from critics. "If X is done, then we are on the road to Nazi Germany" has become a commonplace claim in contemporary bioethical debates.

Sadly, too often those who draw an analogy between current behavior and what the Nazis did do not know what they are talking about. The Nazi analogy is equivalent to dropping a nuclear bomb in ethical battles about science and medicine. Because its misuse diminishes the horror done by Nazi scientists and doctors to their victims, it is ethically incumbent on those who invoke the Nazi analogy to understand what they are claiming.

A key component of Nazi thought was to rid Germany and the lands under German control of those deemed economic drains on the state—the mentally ill, alcoholics, the "feeble-minded," and the demented elderly. They were seen as direct threats to the economic viability of the state, a fear rooted in the bitter economic experience after World War I. The public health of the nation also had to be protected against threats to its genetic health. These were created when people of "inferior" races intermarried with those of Aryan stock. These also included, by their very existence, genetic degenerates—Jews and Roma. Theories of race hygiene had gained prominence in mainstream German scientific and medical circles as early as the 1920s.

What is important to keep in mind about these underlying themes that provided the underpinning for Nazi euthanasia and eugenic practices is that they have little to do with contemporary ethical debates about science, medicine, or technology. Take, for instance, when some congressmen and religious leaders in the United States commented earlier this year that allowing the death of Terri Schiavo, a massively brain-injured patient kept alive by artificial feeding for over a decade, was analogous to what the Nazis had done to Jews in concentration camps. They completely and irresponsibly misused the Nazi analogy. Whatever one thought about the ethical issues raised by the decision to allow removal of a feeding tube from this woman, the decision had nothing to do with the belief that her continued existence posed a threat to the economic integrity of the United States or that her racial background posed a threat to America's genetic health. The fight over her fate was about who best could represent her wishes so that her self-determination could be respected, a moral principle not afforded those killed by deliberate starvation in the Nazi euthanasia programs.

Similarly, when critics charge that allowing embryonic stem cell research permits the taking of innocent life to serve the common good, and then compare it with Nazi research in concentration camps, the claims of resemblance are both deeply flawed and demean the immorality of Nazi practices. Those in concentration camps were used in lethal experiments because they were seen as doomed to die anyway, prisoners, racially threatening, and, given the conditions of total war that prevailed, completely expendable in the service of the national security of the Third Reich.

There are many reasons why a practice or policy in contemporary science or medicine might be judged unethical. But the cavalier use of the Nazi analogy in an attempt to bolster an argument is unethical. Sixty years after the fall of the Third Reich, we owe it to those who suffered and died at the hands of the Nazis to insist that those who invoke the Nazi analogy do so with care.

How the President's Council on Bioethics Lost Its Credibility

A Council Politicized

On March 1, 2004, the White House announced that two members of the President's Council on Bioethics, the widely respected theologian William May and the distinguished biologist Elizabeth Blackburn, were being asked to leave that body. Three new members, a pioneering neurosurgeon and two little-known, conservative political scientists, were taking the vacant places available on the council.

As soon as I read this announcement, I was outraged. The two people leaving the council were among the few who had expressed strong dissent with its majority opinions concerning embryonic stem cell research and the use of cloning to develop stem cell lines for research purposes. I feared that their departure and the selection of their replacements was politically motivated.

The President's Council had, since its inception two years earlier, a distinctively conservative orientation. The previous chair, Dr. Leon Kass, has moved for many years in neoconservative and right-wing circles in Washington. His ambivalence toward new technology in biomedicine is amply documented throughout his many writings over many decades. So, it was not surprising that, when asked by President Bush, he assembled a council, the majority of whom held positions consistent with his own antipathies toward cloning, stem cell research, abortion, prenatal genetic diagnosis, and efforts to slow or reverse human aging.

While this was happening, my friend Gerald Dworkin, a philosophy professor at the University of California at Davis, e-mailed me and asked whether I was aware of what was happening with the membership of the President's Council. I said I was and that I had been thinking about writing a letter of protest to the chair. Gerry suggested that we write a letter to President Bush protesting the changes. I composed a letter and decided to send it out via e-mail to see whether others teaching bioethics wished to sign on.

Within hours, I began receiving responses from all over the country from bioethicists adding their names. More than one hundred signed on within thirty-six hours. Dr. Kass, perhaps knowing that the open letter of protest was circulating on the Internet, published an opinion piece in the *Washington Post*, which appeared on March 3. There he claimed that there had been no purge of dissenters from the council. Moreover, he said that as for the new members, he had no idea where they might stand on issues likely to come before the council since "their personal views on the matters to come before the council are completely unknown."

Kass's claims did not hold up for very long. Blackburn quickly made it known to the media that she considered herself to have been purged. She has spent a good deal of time since her dismissal writing essays for newspapers and science publications decrying Kass's inability to tolerate differences of opinion and blasting the inadequate scientific foundation underlying the council's hostility to advances in biomedicine. Other journalists from the left and the right quickly demonstrated what I already knew to be true—that the new appointees held views that were closely aligned with those of Dr. Kass and the majority of his council. Dr. Kass did not tell the truth in his *Washington Post* article, and the media quickly called him on his dissembling.

The open letter from the bioethics community was made available to the press through the Internet on March 4. By then, nearly two hundred faculty members had signed on, representing all of the major bioethics programs in the United States. Many more would have signed on if the letter had not been restricted to bioethics faculty from the United States. I received dozens of supportive e-mails from students as well as from bioethicists in South Africa, the United Kingdom, Australia, and Germany asking to sign.

It seems clear that this contretemps over the decision to narrow the range of opinion represented on the United States's national bioethics group has fundamentally altered the way bioethics will be done in the future. Never before has the appointment of new members to a bioethical advisory council set off such a heated critical reaction. The decision to turn the council into a council of clones in terms of opinions has gravely imperiled it as a credible body to educate the American public, much less the world, on the ethical issues raised by new developments in biomedicine. Worse still, it is likely that moral debate about the most controversial areas of biomedical research—cloning, stem cell research, the manipulation of the human brain through implants or drugs, and the desirability of trying to slow or reverse human aging will now be ringed with a politically charged corona.

How did bioethics in America reach this regrettable state of affairs?

It is the hysteria over cloning that created the stimulus for the birth of the President's Council on Bioethics. But there was a very special fuel added to this hysteria—the continuing American debate over the morality of abortion that took events in an even stranger direction.

The Council and Abortion Politics

When President Bush gave his first speech to the nation, it was not about terrorism. It was about his opposition to stem cell research. On August 9, 2001, he told the nation that after much prayer and reflection, along with private discussions with two bioethicists, one of whom was Dr. Leon Kass, he had come to the conclusion that destroying embryos for research was wrong. By early April 2002, he had asked Kass to chair the council. But, unwilling to wait for their advice, by the end of the month the president had called for a complete ban on all human cloning both for research and for reproduction. Not surprisingly, the council later that year issued a report that determined after the fact that the president had made the right decision.

The problem with all this banning is that most of it had very little to do with cloning. It has been driven by politics—in this case, the desire to see legislation that grants legal standing to a human embryo. If the Congress would join the president in a ban on cloning embryos for research, since such research involves the destruction of human "life," then those who oppose abortion would have a federal law in place that recognizes the legal standing of the human embryo. Such a law would provide a very useful platform for moving to ban all abortions in the United States. And that is the real goal behind all the Sturm und Drang over cloning from the administration and its supporters in Congress.

Remember the president has had nothing to say about the creation and destruction of embryos in in vitro fertilization clinics. Nor has the Kass council. After some preliminary huffing and puffing, the council issued a report on reproductive technologies that called for a ban on a number of procedures that are of interest to a tiny handful of scientists, and it steered clear of any serious commentary on the practice of destroying or freezing embryos as a part of the effort to help the infertile. Politically, there is no support for interfering with infertility treatments. They may involve the destruction of embryos, but the treatments are manifestly prolife, and President Bush and his advisers are well aware of the popularity that infertility treatment enjoys among the American people. So it is human cloning for research that has been the punching bag for those keen to limit abortion rights in the United States.

What to Do Now?

At this point, the council has lost much of its credibility as a fair and balanced body that can engage the American people in a full-fledged debate about cloning or, for that matter, any other biomedical advance. The council still retains its status as a powerful body that has the ear of the president and the administration as long as it does not cross any dangerous political lines, such as the one that infertility treatment represents.

There is no turning back from the politicization of bioethics in the United States. Bush cannot alienate his conservative base, and opposition to abortion is a key litmus test for that base. It is unlikely that the President's Council can serve as a credible, fair broker of bioethical dialogue for all Americans. As is true now of so much in American political life, publicly sponsored bioethics has now become just another weapon in the arsenal of partisan politics.

Pray It Ain't So

WOULD PRAYING for infertile couples undergoing infertility treatment help them get pregnant? A study published in the *Journal of Reproduction* three years ago seemed to show that prayer made a difference. A big difference.

The study claimed that women who, unknown to them, had strangers praying for them got pregnant at double the rate of those who did not! Since one of the authors of the study was the chair of the department of obstetrics and gynecology at Columbia University and another a prominent California fertility specialist, it was hard to deny that real empirical evidence had been found for the power of prayer.

But it was the third author of the study, Daniel Wirth, a self-styled psychic researcher, who set off alarm bells in some of those who read the study. Wirth is currently under arrest in California having pled guilty to bank and mail fraud charges.

The revelation of Wirth's naughty behavior led Dr. Bruce Flamm, the director of research at Kaiser Permanente Hospital in Riverside, California, to go back and reexamine the prayer-boosts-pregnancies study. His conclusion: the whole study is probably completely fabricated. The design of the study is, he thinks, completely unscientific.

Fraud and hanky-panky aside—and that is admittedly no small aside—there is something fundamentally wrong with this and the many other studies of the power of prayer to effect healing that pop up from time to time.

The problem is not that every study of prayer is fabricated. The real problem is that such studies mix standards of evidence that ought not go together at all.

The proof that something works in medicine and science comes from the careful, systematic observation of the facts. One population is exposed to a treatment and studied in comparison with another similar population that is not. The data gathered show the power of the treatment. Prayer does not fit into this kind of research framework.

The power of prayer comes from faith. It is not something you can refute with a study or a controlled trial. Anyone who believed in the power of prayer who stopped praying because the data came in negative might reasonably be said not to really believe in prayer.

The problem with any study that tries to see whether prayer works is that it can never prove any such thing. Those who believe in prayer will continue to believe, and those who don't won't. The scientific study of prayer has no more chance of proving prayer ineffective than the scientific study of geology does proving that the biblical account of creation is just a story. Those who have faith will still believe. And from a scientific point of view, that is just fine.

Who Wins When Religion Squares Off against Science?

WHICH SIDE should win when religion squares off against science? The media have treated this as the crucial question being put front and center by the renewed effort to push the teaching of religion as science in the guise of intelligent design (ID). Proponents of ID have evolution in their sites in biology classes throughout America. The question, according to the media is, Who is going to win? But this is the wrong question. The real question is, How is it possible that science finds itself stuck in the middle of this debate in twenty-first-century America in the first place?

Some members of the scientific community do not want to engage the proponents of ID. They maintain that to do so is to give credit to a point of view that is patently wrong. But they are completely and utterly wrong.

It is absolutely essential to engage proponents of intelligent design. Not only to ensure that high school kids in America graduate with some knowledge of one of the key foundational theories of modern biology but because what is really at stake is educating every American about the nature of science and scientific discourse. And as recent events well beyond the scope of the debate over ID quickly reveal, no one should take for granted the idea that Americans know what they are talking about when it comes to distinguishing science from either nonscience or nonsense.

Why is it so difficult for some Americans to see that a position that holds that the creatures and plants around us are so unique and complex in their design that the only way to explain their existence is by invoking a grand designer (otherwise known as God) is a religious explanation? The main reason is that too many Americans have no idea what makes an explanation a part of science as opposed to a part of religion, fiction, or fantasy.

Critics of evolutionary theory in making the case for ID delight in pointing out that there are phenomena that are difficult to explain by means of natural selection acting on genetic differences over long periods of time. And they are partly correct. There are many biological facts that evolution has had a difficult time explaining. But, ironically, that is one of the distinctive characteristics of a scientific theory—that there are facts or phenomena that seem inconsistent or at odds with the validity of the theory. What is presented as a vice, a limit on the explanatory power of a theory, is in reality a defining virtue of science.

Consider the case of social behavior among sterile insects. For decades biologists knew that ants, bees, termites, and many other insects had members with highly developed behaviors but that were completely sterile. How could a theory that explains complex behavior by the slow process of selection acting on small genetic changes manifest over long periods of time possibly account for these critters? Darwin's critics had delighted in confronting him with the reality of sterile social insects. Any fan of intelligent design would have drooled at the prospect of torturing later generations of evolutionary biologists to try and explain what was obviously and certainly inexplicable—if they had known enough biology to do so. They didn't. But biologists still realized the nature of the problem. There was a limit to what evolutionary theory could explain. Except that there wasn't!

It took some hard thinking by some of the giants of late twentieth-century evolutionary biology, such as W. D. Hamilton, R. L. Trivers, and E. O. Wilson, but biologists recognized that evolution does not have to act on individuals to select for social traits—it acts on genes. And if animals or plants have enough genes in common, then evolution could select for very complex behavior in sterile insects if that behavior conferred enough advantage on other animals or plants with the same genes—in the case of the social insects, queens and other unique egg layers. Kin selection and reciprocal altruism could explain what up until the 1970s had seemed inexplicable. By modifying their understanding of how evolutionary forces worked from individual organisms to genes, biologists were able to see how some environments could elicit the existence of sterile social insects.

There is a very clear moral to this story. What makes something scientific is not, as proponents of ID would have it, that a theory can explain everything. It is that it is possible to imagine or even encounter a fact that will or does imperil the truth of the theory.

What makes an explanation scientific is not that it is true. It is, rather, that an explanation that seems valid or correct can be tested, verified, or falsified. Truth—or at least the blind allegiance to the validity of a doctrine or explanation, come hell or high water—is of necessity a matter of faith. There are certainly scientists who push on with their pet theory or hypothesis in the face of the evidence. But at some point, be it astrology, Marxist explanations of history, Freudianism, UFO abductions, leeching, or alchemy, proponents either concede that they could be wrong or get consigned to the categories of ideology or faith.

Proponents of ID do not admit that they might be wrong. There is nothing that could possibly show why the invocation of a designer or God as the explanation for the diversity of life on Earth is false. That does not make ID false. But it surely does exclude it as science.

So if it is so clear that ID is religion gussied up as science, then why it is that science has fallen into such disrepute throughout many sectors of American society? And this dismissal of science is hardly confined to evolution.

Science and religion subtly squared off over the subject of why so many intense hurricanes have devastating the Gulf region in recent years. The Reverend Franklin Graham, son of the Reverend Billy Graham, thinks New Orleans was targeted because of the city's sinful reputation. At a speech in Virginia, he said, "This is one wicked city, OK? It's known for Mardi Gras, for Satan worship. It's known for sex perversion. It's known for every type of drugs and alcohol and the orgies and all of these things that go on down there in New Orleans. . . . There's been a black spiritual cloud over New Orleans for years."

Jennifer Giroux, director of Women Influencing the Nation, a fundamentalist group, said on national television that she did not "have any problem with what the Reverend Graham said. I believe that, when he talks about New Orleans and a cloud of darkness hanging over an immoral city, he could be referring to any city in the United States. I believe that God will not be mocked. . . . But let's look at the state at which our country is in, abortion, contraception, homosexuality, cloning, I mean, it all really comes together in one big picture."

The ninety-two-year-old archbishop of New Orleans also ventured into the explanatory arena concerning the weather. He claims that the hurricanes that hit New Orleans are a chastisement of the city and a chastisement of the United States.

The media are flush with many more such claims that bad weather is God's vengeance on a sinful nation, group, city, or person.

Why does this drivel proliferate? Why do the claims of those professing that babies and the frail elderly are being killed by a wrathful God in New Orleans for the sins of out-of-town gamblers and revelers or that kids on break from college are being drowned by God to show that adherence to Islam is the only true path go unchallenged when science has sound explanations to offer for the weather, including hurricanes, tornadoes, typhoons, and tsunamis?

It is because our culture does not understand the difference between faith and science. The media often present them as two sides of a story in the name of balanced reporting. Those who fear science simply ignore scientific explanations or dismiss them as inconsistent with religious accounts. Worse still, scientists do not even enter the fray, either believing themselves to be above such mundane matters or uncertain themselves about what are the defining characteristics of science.

If the United States is going to deal with the challenges of global warming, air and water pollution, emergency contraception, vaccination, species extinction, waste disposal, abortion, cloning, and stem cell research, among other crucial problems, then scientists must be engaged about topics when religious accounts appear on their turf. And much more attention needs to be paid to what it is that separates science from faith. Both can most assuredly coexist. But that is precisely because they are very distinct.

Why Are These Nuts Testifying?

THE MOST SURPRISING thing about the recent testimony
by Severino Antinori, Panayiotis, Zavos, and Brigitte Boisselier
before the National Academy of Sciences in March 2001 is not
that they announced that they intend to create a human clone. Nor is it
that they faced near-universal criticism from virtually everyone in the dis-
parate fields involved: embryology, animal cloning, human reproduction,
and bioethics. The most surprising thing about the testimony was that it
took place at all. The National Academy of Sciences (NAS) is one of the
most important, august scientific bodies in the United States. When they
decided to hold a hearing on cloning, it received attention: from the
media, from scientists, from Congress. How did it happen that this rep-
utable bastion of American science welcomed as equals cooks and cranks?

Brigitte Boisselier runs a lab for Clonaid, a front for the Raelian religious movement. They believe that humans are a result of alien cloning,
and therefore cloning technology brings us closer to the aliens who created us. Zavos left behind a trail of issues and problems that led to his
"extrication" from his affiliation with the University of Kentucky and a
successful lawsuit against him. Antinori is infamous for pushing the envelope ethically in his reproductive work and is likely to soon lose his medical license (neither Zavos nor Boisselier are clinicians—hence they have
no license to practice medicine). None of the three has produced any
prominent peer-reviewed published research that bears on the safety
questions that cloning humans raises. None of the three has any standing in the relevant scientific fields at stake in cloning. None of the three
has been engaged in the kind of animal experimentation that would
allow them to perfect the delicate techniques cloning requires to be successful. Each of these three is lodged firmly on the far fringes of science.

When the NAS provides a visible platform to individuals such as
these, it serves to legitimize them and erodes the distinction between science and pseudoscience. At a time when many groups are attempting to

legitimize pseudoscience, from "scientific creationists" to "alien abduction experts," it is imperative that our leading institutions make clear the differences between good and bad science. The academy failed in this mission when it let the kooks of cloning sit as equals.

There is no doubt that the testimony of the experts who were present at the panel will lead to a resounding condemnation of the practices of Zavos, Boisselier, and Antinori. And the NAS undoubtedly thought it important to hear from them on the one-in-a-million chance that they had actual, plausible data that would succeed in overturning all of the research by the mainstream scientific community. But the decision to treat these people as scientific equals was wrong. Will the NAS invite "creation scientists" if they do work on evolution? Are they readying invitations to those "scientists" who believe in the healing power of crystals for their next report on cancer? What about astrologists?

By providing this forum to a group with no scientific credentials, the NAS has invited the public and the media to treat this issue as a debate among (equal) scientific authorities. Nothing could be further from the truth. The attempt to clone a human at the present times is bad science, bad medicine, and unethical—and no one should be fooled into thinking that there is a serious difference of opinion among serious scientists about this fact.

Part XI

Donation and Transplantation of Organs

About Face

THERE ARE TWO basic types of full facial transplantation. Both require that the donor's face be harvested within six hours of his or her death. The more extensive approach involves the transplantation of the donor's bone structure and soft tissues. While more complicated, using this strategy, the recipient would be more likely to have an aesthetic result, albeit closely resembling the donor. A less extreme approach is to transplant the donor's soft tissue onto the bone structure of the recipient. This procedure should produce a "hybrid" in which the features of the face resemble both donor and recipient.

The first candidates for such surgery will be those who suffer disfigurement as a result of accident or disease. Current surgical techniques for facial reconstruction have had poor outcomes. Patients often undergo dozens of procedures and still remain disfigured with functional limits—faces that are unable to communicate the subtleties of human expression. Many surgeons believe that transplants offer the facially disfigured the best chance of a life where they can open their eyes, eat, speak, and go out in public. But the ethical questions raised by the prospect of facial transplants have received inadequate attention. They are enormous.

The most obvious is the high risk of failure. Facial transplantation is completely experimental since it is currently based on few studies involving animals. Unlike the transplant of, say, a pancreas, a face transplant requires not only sufficient blood supply but also the reconnection of intricate nerve endings to ensure functionality. However, hand transplantation, which is the closest type of transplant to what is being proposed, has fallen far short of expectations with respect to the reconnection process. Little functionality has been obtained with hand transplants. Face transplants are an even greater challenge since the recipient's nerves may be damaged from whatever caused the original disfigurement.

Face transplants will require the provision of constant immunosuppresion. While the risks associated with immunosuppressant drugs are seen as acceptable when the alternative is certain death, face transplants would not be undertaken for that purpose. Once a face is transplanted, the procedure would be irreversible; and if the face rejects it, it would be, simply stated, a gruesome horror.

Who would be selected to take on these risks? The most disfigured, those recently disfigured, those with the least function, those who are the unhappiest with their appearance, children or adults, the terminally ill, or those with no family or friends should failure ensue? The question of who would be the first subject is critical since after the risk of failure, the next major ethical challenge of face transplants is the threat to the personal identity of the recipient.

Most transplants are not readily observable by others. Face transplants are the most observable form of transplantation imaginable. While it is true that studies show that many of the disfigured define their identity by who they are on the inside and often feel disconnected with their external visage, the fact remains that it will be challenging, to say the least, to adjust to the presence of the tissues most closely associated with each individual's personal identity.

The issues raised by donation are also unprecedented. It is not clear who should be approached to donate—the terminally ill, anyone with a signed donor card, those who have no known criminal or psychiatric history, those with families, or those without them? All most proponents have said is that it is essential to match gender, race, and age within ten years.

The very idea of relinquishing one's own face or the face of a loved one will for practical purposes pin the needle on what I referred to more than a decade ago as the "yuk" factor. Undertakers, as well as those who request autopsies and organ donations from families of the deceased, overwhelmingly report that it is very important to people that the appearance of the deceased's face be preserved. A family member or partner may feel guilty burying a faceless corpse, especially when American death rituals often involve open caskets. And no one has any idea what impact seeing the face of their deceased loved one on someone else's body would have on those who were close to the deceased donor. While cadaver donation almost always is done without the disclosure of the donor's identity, the identity of a recipient would be very difficult to conceal from a donor's relatives or friends.

Surely so momentous an operation ought not to be left to the whim of any particular surgeon or surgical team. That is precisely what happened with the first human hand transplants, leading to a great deal of controversy and public concern. That is a history that we should not be asked to face again.

Restricting Blood Donations or Mad Cow the Deadlier Threat?

THE AMERICAN RED CROSS and the Food and Drug Administration (FDA) think it is time to do something to prevent mad cow disease from killing you. That is certainly commendable. But the thinking behind the Red Cross's proposal—to prohibit anyone who has been in any of the Western European countries where the disease has appeared from donating blood—might make less sense than you may think.

The number of countries reporting cases of deadly mad cow disease has been growing in recent weeks. In the past few days alone, the human death toll in the United Kingdom from the human version of mad cow disease reached eighty, Italy reported its first confirmed case of an infected cow, and Spain detected two new cases in cattle. There are reports of hundreds of stricken cows in France, Ireland, and Portugal and infected people in France and Ireland.

People get the disease by eating meat from animals that have mad cow disease. The animals get the disease by eating feed made from parts of other animals that are infected.

The FDA and the Red Cross, as are many officials in Europe, are worried that someone unknowingly infected with the tiny viruslike bug known as a prion that causes the human version of the disease will spread it to others if the person donates his or her blood.

Since there is no effective test to detect this microscopic killer, one way to keep it out of the blood supply is to prohibit anyone who has visited countries where the disease has been reported from donating. The FDA has already banned blood donations by anyone who has spent just six months or more in Britain, and an advisory panel recommended Thursday widening the ban to include long-term residents of France, Ireland, and Portugal.

The Red Cross has indicated that it may go even further and extend the ban to all people who have spent time in Western Europe.

Well, why not? Why shouldn't the Red Cross tell those who have visited France, spent time as a student in Italy, or served in the military in the United Kingdom that their blood is no longer useful?

The reason is that there is no confirmed case of anyone ever getting the human form of mad cow disease—known as Creuzfeldt-Jakob disease—from a blood transfusion. There is only one case of a sheep dying from the disease after receiving the blood of another sheep. No human cases and one animal case means that the jury is still out on blood transfusion as a way to spread mad cow.

What is very certain is that if everyone who has ever passed through Heathrow or Gatwick, munched a croissant in front of Notre Dame, or quaffed a couple of beers during Oktoberfest is told not to donate, there will be a lot less blood around in our blood banks. And that is certain to hurt and even kill people.

The nation has been operating right on the edge of having enough blood for all who need it for the past few years. A drop of even a few percentage points in donors—which a ban on donations from Americans who have been in affected Western European countries since 1980 would certainly bring about—would mean there may be no blood when you need your emergency bypass operation, unexpected Cesarean delivery, or transfusion due to a burn, car accident, or other trauma.

Keeping donors out of the blood supply gives a very small measure of protection against a theoretical risk. Keeping donors out of a blood supply that is growing increasingly scarce every year means that there may be no blood at all in the blood bank when you need it.

That is the stark choice we face in trying to decide what is the best course of action to ensure that mad cow disease does not wind up killing people because we have no blood to give them.

Jumping the Line

LAST YEAR, thirty-two-year-old Todd Krampitz was battling cancer. His liver, riddled with a huge tumor, was starting to fail. Then a courageous family in another state who had tragically lost a loved one made the vitally important decision to donate a liver. On August 14, Krampitz got a new liver. So, the system worked and a young man's life was saved. Except that in this case the system did not work. The fact that Todd Krampitz got a liver is unethical.

There is a very long waiting list in the United States of people waiting for livers. The list they wait on is run by the United Network for Organ Sharing, a quasi-public agency, based in Richmond, Virginia, that operates with a grant from the Department of Health and Human Services. The national list was created so that everyone who needs a life-saving organ has a fair chance at getting a transplant. At the time Todd Krampitz got his liver, more than seventeen thousand people on that list also needed a liver transplant, some more desperately than he did. More than a thousand of those live in Krampitz's home state of Texas. So what was wrong with Krampitz getting a liver? He got his by cutting in line.

Krampitz and his doctor knew about the national system for distributing organs. But he and his wife decided to try to circumvent the system. They took out billboards around Houston, bought newspapers ads, and appeared on various national television shows. Their primary goal was not to get Americans to donate their organs. Rather, it was to get a family to donate a liver directly to Todd. That is what happened. A family in another state donated a liver specifically for Todd Krampitz. The fact that they could is wrong.

Todd Krampitz has the right to try to do whatever he can to save his life. But the whole point of the organ distribution system, which has been in place since 1986, is to give everyone in need of a transplant, not just those who can pay for billboards and grab national media attention, an equal shot at the scarce supply.

The system that has worked so well for almost twenty years uses a complex set of formulas involving blood type, tissues type, size of donor, medical urgency, and likelihood of survival to distribute organs. Todd Krampitz would not have gotten a liver using that formula. He was not as near death as others were on the day of his operation. And his cancer is so advanced that it may be impossible to make a transplant work for him. At least one experienced Texas transplant surgeon declared that it was a "sad day for liver transplantation" when a person not at the top of the waiting list could hijack an organ.

No one, including me, wants to begrudge anyone the right to do what they need to do to save their lives. But in the case of organ transplantation, the organs needed to save lives have to come from all over America from people from all sorts of backgrounds and financial means. We have a system in place that makes sure that these precious gifts are given out so that everyone who needs an organ has a fair chance to get one. Todd Krampitz found a loophole in that system and used it. In the end, someone else who was not on a billboard or national television died because Krampitz was not told to wait his turn. Congress, the head of Health and Human Services, and the president should act quickly to close that loophole so that individual self-interest cannot destroy the common good.

www.matchingdonors.com

O N OCTOBER 20, 2004, surgeons in Denver, Colorado, removed a kidney from Robert Smitty and transplanted the organ into Bob Hickey, a fifty-eight-year-old physician suffering from renal failure. Hickey had paid MatchingDonors.com, a commercial website, a $295 monthly fee to advertise his need for a kidney on its site.

Smitty, a thirty-two-year-old part-time photographer from Chattanooga, Tennessee, found Hickey on the MatchingDonors.com website. He is the first person in the United States known to have arranged to make an organ available to a stranger using a commercial company as a middleman.

On October 28, 2004, Robert Smitty became the first person involved in an arrangement brokered by a commercial website to land in jail. Smitty had failed to pay his child support, and, since his face was all over the newspapers, he was jailed on his return to Tennessee. Suspicion that he had sold, not donated, his kidney remains rampant.

The story of Robert Smitty's kidney highlights the need that exists to find ways to allow those waiting for transplants to locate living persons who might want to donate an organ to a complete stranger. At present, unlike cadaver donation, there is no national system for bringing donors and recipients together. Smitty's story also highlights the absolute lack of any oversight of this new form of organ donation.

In 1997, there were no documented instances of absolute strangers donating organs. By 2003, there were dozens of such cases. The numbers continue to grow. So do the number of sites on the Web "brokering" organs. So what is wrong with a practice that gives those who need organs a chance to find them?

A key ethical issue in soliciting strangers by means of advertising on the Web is that the practice is unfair. Those who can pay middlemen to publicize their plight have greater access to potential lifesaving transplants.

And the prospects for coercion and extortion are, in the wild and wooly world of the Web, staggering. As Mr. Smitty reminds us, for every truly generous person who wishes to help a complete stranger, there are plenty who see the redundancy of their kidneys as a quick ticket out of debt or the fast route to the good life. Unregulated brokerage in organs on the Internet means organ sales will flourish.

Still, there are those who would defend the rights of private citizens to communicate with one another in order to broker whatever arrangements they wish with respect to their body parts. But, even if the line between altruism and commerce is allowed to grow fuzzy, other ethical challenges are yet to be addressed.

Should those who want to help by giving up a body part have to undergo a psychological evaluation first? No rules govern how long a stranger who wants to donate should have to wait, who will pay if a donor dies or is injured, or what medical criteria should govern donor eligibility.

Perhaps most disturbingly, there is no consensus among transplant centers as to the degree of contact that should be permitted between those making organs available and those who need them either before or after surgery.

There is no, in principle, ethical reason I can think of why a healthy, informed person cannot rationally decide to give another person a kidney. But there are many reasons why the emerging practice of commercially brokered Internet matching between strangers should be viewed with a good deal of ethical skepticism as to whether donor or recipient interests are currently adequately protected.

Misguided Effort to Ease the Organ Shortage

THE GOVERNMENT'S goal of boosting organ donation is a good one. Unfortunately, though, the plan put forward by Tommy Thompson, the secretary of Health and Human Services (HHS), will not work.

Thompson has a problem. His predecessor, Donna Shalala, instituted a policy that would require transplant centers to share organs needed for transplants regionally and nationally. With organs in such short supply, Shalala decided that the fairest way to distribute them was to give them to the person most in need regardless of where they lived or worked.

Before becoming the head of HHS, Thompson was the governor of Wisconsin. In that job, he led the fight to resist organ sharing. In fact, he signed a law making it clear that organs donated in Wisconsin would have to stay in Wisconsin regardless of whether or not people in other states might have a more pressing need for those organs to stay alive. The problem Thompson now faces is what to do with respect to the policy of sharing organs by need, not geography—implement it, or try to get it rescinded.

So far, Thompson has decided to punt. He announced at a press conference Tuesday that his goal was to get rid of the problem of what is the best way to distribute scarce organs by making sure that this nation has enough organs to meet the demand for them.

"Let's work together," he urged the nation. "Why can't we solve the problem [of who gets scarce organs] instead of creating more angst among ourselves?"

Even if Thompson were to somehow double or even triple the number of organs that Americans donate, he would still have to live with some angst. The issue of who lives and who dies is not going to go away any time soon. The waiting lists for transplants are long, and, in a graying society like ours, they grow longer every day.

Right Idea, Wrong Approach

Put aside the fact that Thompson is going to have to decide whether he favors states' rights over saving lives. Getting more organs is an ethically sound idea, but his new plan won't work.

The ideas Thompson presented Tuesday about how to get more organs involve a partnership with private businesses to promote organ donation among employees, a medal to honor donor families, more teaching about organ donation in teenage driver education classes, and a national donor card that makes it clear that anyone who has a card can serve as a donor even if his or her family members object.

I wish there was some reason for optimism about these initiatives, but there isn't. Americans are well aware of organ donation. Public education campaigns have been letting people know about the "gift of life" for thirty years. One more pamphlet at the office or a few more minutes in driver's ed is not going to boost the percentage of people who sign donor cards beyond what has already been achieved through concerted public education efforts.

Moreover, as study after study has shown, signing a donor card is not a very effective way to ensure organ donation since cards are often lost, not available when someone dies, or simply not found by medical personnel desperate to try to save a life.

Nor will the ideas of medals and tougher donor cards do much good.

No Heroics

These initiatives send precisely the wrong message.

Giving medals to the families of those who give organs makes donation an act of heroism. It isn't. Acting as an organ donor is something that everyone should simply be expected to do because it is the right, the humane, and the decent thing to do with your body when you die.

As tempting as it may seem, it isn't even prudent to push for policies that tell families their views and feelings do not count when their loved one dies. Does anyone at HHS really think—really—that doctors and nurses will ignore the protests of a young woman whose husband has just died in a car accident and who does not want organs taken by pointing to his donor card and then removing his heart?

There are some steps that are likely to produce more organs to save lives. First, remove financial barriers in access to transplants.

Poor people do not donate their organs at the same rate as the rich. The reason is simple. Poor people, many without health insurance, know that the rich have a much better chance of getting transplants than they do. If Thompson really wants to get more organs available for transplant, he must make sure that the rich and the poor have the same chance to get a transplant.

Second, treat organ donation as expected, not heroic. Thompson and President Bush should appear on national television with every governor and member of Congress. Every one of them should fill out a donor card. And each should do it simply because it is the right thing to do. Shortly thereafter, every high school principal, religious leader, CEO of a major company, and media celebrity should do the exact same thing on the exact same day.

Every doctor and nurse should approach organ donation as the right thing to do and expect that it is what people will want to do. If there are objections, then the burden should be on individuals and families to raise them.

Shortage is always going to be a concern with organ transplantation. But we can get more organs if the public believes that everyone has a fair chance to receive a transplant should they need one and that donation is not a matter of heroism but of common decency.

No Excuse for Blood Donor Bias

THE ONGOING threats of terror on American soil mean that the chance of not having enough blood on hand is a daily risk that every one of us faces. But we are overlooking a much-needed source of donor blood: men who have had sex with men. It's time for the government to reexamine its policy of automatically excluding them.

If you need a blood transfusion or a blood product, there is a very real danger that it might not be available. At different times during the year, blood banks and hospitals are unable to meet the demand for blood.

And the shortage is growing worse as more Americans undergo bypass operations, organ transplants, Cesarean sections, hip and joint replacements, and other procedures.

One potential source of blood we are not exploiting, however, is men who have ever had sexual relations with other men. Under a policy set by the Food and Drug Administration (FDA) seventeen years ago, any man who has had sex with another man one or more times since 1977 is automatically disqualified as a blood donor.

The policy of banning all of these men as possible donors in the face of this nation's worsening blood shortage makes little sense.

There is no reason to presume that those who engage in dangerous heterosexual practices that might lead to infection and spread of disease are any less a source of risk than those who engage in male-to-male sex. Yet, the FDA has not permanently banned heterosexuals who engage in unsafe sex as prospective blood donors.

Advances Address Concerns

The rationale for changing the policy recently became even more persuasive. A technological breakthrough should remove any lingering concerns about the safety of allowing some gay or bisexual men, or men who simply experimented with male-to-male sex once in their lives, to donate blood.

The breakthrough: FDA approval of a new screening technology for finding HIV and hepatitis C in blood. This technology, brought to us by Chiron, a biotech company in Emeryville, California, screens for the presence of viral DNA.

Whereas there used to be a period of about three months after infection with HIV or hepatitis when a person tested negative, the new screen can detect the presence of either virus almost immediately and with uncanny accuracy.

This advance comes quick on the heels of a new study that shows that screening people for risky behaviors with respect to hepatitis C is no substitute for testing their blood. Risk screening relies on the truthfulness of donors in revealing their sexual and medical histories—if they even know their history.

Clearly, the testing of blood is the only reliable measure. So if a man or woman tests negative for the AIDS virus or any other communicable disease, he or she should be allowed to donate.

The policy of forever excluding people who have ever had male-to-male sex should have been changed years ago. Now there's no excuse. The question is whether the FDA and Congress will act or simply let old prejudices about risk stand in the way of finding answers to the very real risk that the growing shortage of blood poses for every American.

Sperm Transplants Should Spur Debate

N THE LATEST scientific breakthrough to spark a reaction on the ethical concern meter, scientists at the University of Pennsylvania's School of Veterinary Medicine have performed the first successful transplant of testicular tissue from one species to another.

They took sperm-generating tissue from different species and put it on the back of immunosuppressed mice. The tissue took root and began to churn out sperm. Put simply, the veterinary researchers at Penn have made it possible for a mouse to produce sperm from pigs and goats. There is no reason to think that it would not be possible for other species, including humans, to have their testicular tissue transplanted into this type of mice as well.

There are some good reasons for performing this sort of transplant. For example, if there is a species that is nearing extinction, say, the cheetah, it would be possible to transplant tissue from a cheetah to a mouse and then have an inexhaustible source of sperm to use to restore that species.

Plus, a mouse that can make sperm will help scientists study the process of sperm formation so that it should be possible to figure out what diseases and disorders cause infertility in animals—studies that can be rather difficult to do without the risk of getting eaten or trampled by your subject.

Ethical Questions

But there are obvious ethical questions about this experiment. Whether it is right to transplant reproductive cells across species is one such question. Some might say it is unnatural to have a mouse that can make the sperm of a goat or a pig. However, this argument is not compelling when you consider that almost everything that goes on in medicine, agriculture, and animal breeding involves something that is "unnatural." The ethical question is whether there is sufficient benefit in doing something unnatural to let it be done.

The more troubling ethical questions arise when this breakthrough in transplantation moves toward human application. Should there be a limit placed on using other animals as hosts of human sperm? Would we feel comfortable having scientists use mouse models to study human infertility or to develop better forms of male contraception?

And what about the man who faces sterility due to cancer or some form of medical treatment requiring radiation? Should he be allowed to store his sperm-making cells inside a mouse so that he could have children in the future? It's not impossible to imagine people trying to store their sperm for eternity on the back of a mouse as a way to cheat death.

It is too soon, in my view, to answer any of these questions. It is also too soon to know whether you could reliably make healthy offspring from transplanted sperm cells. But it is not too soon to ask these questions in order to start the process of finding answers. Consider yourself asked.

The Return of Fetal Tissue Transplants

A **DECADE AGO,** several years before a sheep named Dolly was cloned, and long before stem cells became a significant source of debate, the most controversial bioethics issue of the day was the question "Is it permissible to use fetal tissue for research?" At that time, there was hope that fetal brain tissue could be transplanted to people with many serious diseases, including Parkinson's disease and diabetes. Much of that research failed to bear fruit, and the focus of scientific hope and bioethical controversy moved on to today's battles over cloning and embryonic stem cell research.

However, a major breakthrough by researchers at Stanford University should take the debate about embryonic stem cell research back to the future of fetal tissue transplantation.

In a paper published in the *Proceedings of the National Academy of Sciences*, the Stanford group reported finding that fetal brain stem cells may be capable of growing into neurons and thereby fill in the gaps in the brain that are often associated with debilitating strokes. Researchers injected stem cells derived from fetal tissue into the brains of rats. These cells migrated to the injured location and turned into the appropriate kinds of neurons, thereby repairing much of the damage from the stroke.

More than seven hundred thousand Americans suffer a stroke every year. For many patients and their families, the consequences are devastating. The possibility that we may be able to prevent the transformation of vital, active individuals into people with little or no functional ability is one that would seem to demand our full support.

While the research at Stanford merely shows success in rats, this is, nonetheless, a significant step in the fight against strokes—if the political will is there to push the research forward.

What makes this research especially controversial is the fact that the cells to be transplanted are derived from aborted fetal tissue. While embryonic stem cell research faces a variety of obstacles, fetal stem cell research, though legal and even receiving some federal support, places this research squarely in the path of those such as President Bush who so strongly oppose embryo research. When abortion is mentioned, those who oppose it tend to oppose anything connected with it.

Does this make sense? Is the use of tissue that would otherwise be incinerated or destroyed unethical? We often find ways to derive benefit from tragic circumstances. Young trauma patients who are injured in car or motorcycle accidents, or who are victims of gunshot wounds, are leading sources for organs used in transplants. Simply because we choose to create some good from a terrible event does not mean we must support or approve of the event.

Abortion is a legally approved medical option in the United States. The fetal tissue that results could potentially help save lives. There is no safe source for fetal tissue for clinical research outside aborted fetuses.

The key to ethically using fetal stem cells is to make sure that we have adequate safeguards in place to ensure that high standards are met in the way this research is conducted. The most important safeguard would be to never ask about obtaining fetal tissue until after an abortion has been performed.

There are hundreds of thousands of abortions conducted in this country and millions more in others every year. We are left with weighing the potential benefit to millions of people against using tissues that will otherwise be destroyed.

If you or your family is to benefit from what the Stanford researchers have found, then we need a political consensus that not only recognizes the importance of stem cell research but also is willing to accept some use of fetal tissues.

What Is Bioethics?

The Nature and Scope of Bioethics

Bioethics as a field is relatively new, emerging only in the late 1960s, though many of the questions it addresses are as old as medicine itself. When Hippocrates wrote his now famous dictum *Primum non nocere*— "First, do no harm"—he was grappling with one of the core issues still facing human medicine today—namely, the role and duty of the physician. With the advent of late twentieth-century science, an academic field emerged to reflect not only on the important and age-old issues raised by the practice of medicine but on the ethical problems generated by rapid progress in technology and science. Forty years after the emergence of this field, bioethics now reflects the profound changes in medicine and the life sciences.

Against the backdrop of advances in the life sciences, the field of bioethics has a threefold mission: to raise important questions about the general practice of medicine and the institutions of health care in the United States and other economically advanced nations, to wrestle with the novel bioethical dilemmas constantly being generated by new biomedical technologies, and to challenge the presumptions of international and population-based efforts in public health and the delivery of health to economically underdeveloped parts of the globe. While attention to the ethical dilemmas accompanying the appearance of new technologies such as stem cell research or nanotechnology can command much of the popular attention devoted to the field, the other missions are of equal importance.

At the core of bioethics are questions about medical professionalism, such as "What are the obligations of physicians to their patients?" and "What are the virtues of the 'good' doctor?" Bioethics explores critical issues in clinical and research medicine—including truth telling, informed consent, confidentiality, end-of-life care, conflicts of interest, nonabandonment, euthanasia, substituted judgment, rationing and access to health care—and in the withdrawal and withholding of care. Only minimally affected by advances in technology and science, these core bioethical concerns remain the so-called bread-and-butter issues of the field.

The second mission of bioethics is to enable ethical reflection to keep pace with scientific and medical breakthroughs. With each new technology or medical breakthrough, the public finds itself in uncharted ethical terrain that it does not know how to navigate. In what is very likely to be the "century of biology," there will be in the twenty-first century a constant stream of moral quandaries as our scientific reach exceeds our ethical grasp. As a response to these monumental strides in science and technology, the scope of bioethics has expanded to now include the ethical questions raised by the human genome project, stem cell research, artificial reproductive technologies, the genetic engineering of plants and animals, the synthesis of new life forms, the possibility of successful reproductive cloning, preimplantation genetic diagnosis, nanotechnology, and xenotransplantation, to name only some of the key recent advances.

Bioethics has also more recently begun to engage with the challenges posed by delivering care in underdeveloped nations. Whose moral standards should govern the conduct of research to find therapies or preventive vaccines useful against malaria, HIV, or Ebola—local standards or Western principles? And to what extent is manipulation or even coercion justified in pursuing goals such as the reduction of risks to health care in children or the advancement of national security? This population-based focus raises new sorts of ethical challenges both for health care providers who seek to improve overall health indicators in populations and for researchers who are trying to conduct research against fatal diseases that are at epidemic levels in some parts of the world.

As no realm of academic or public life remains untouched by pressing bioethical issues, the field of bioethics has broadened to include representation from scholars in disciplines as diverse as philosophy, religion, medicine, law, social science, public policy, disability studies, nursing, and literature.

The History of Bioethics

Bioethics as a distinct field of academic study has only existed for forty years, and its history can be traced back to a cluster of scientific and cultural developments in the United States in the 1960s. The catalyst for the creation of the interdisciplinary field of bioethics was the extraordinary advances in American medicine coupled with radical cultural changes during this time period. Organ transplantation, kidney dialysis, respirators, and intensive care units made possible a level of medical care never before attainable, but these breakthroughs also raised daunting ethical dilemmas that the public had never previously been forced to face, such as when to initiate admission to an ICU or when treatments such as dialysis could be withdrawn. The advent of the contraceptive pill and safe techniques for performing abortions added to the ethical quandaries of the "new medicine." At the same time, cultural changes placed a new emphasis on individual autonomy and rights, setting the stage for greater public involvement and control over medical care and treatment. Public debates about abortion, contraceptive freedom, and patient rights were gaining momentum. In response, academics began to write about these thorny issues, and scholars were beginning to view these "applied ethics" questions as the purview of philosophy and theology. "Bioethics"—or, at the time, "medical ethics"—had become a legitimate area of scholarly attention.

In its early years, the study of bioethical questions was undertaken by a handful of scholars whose academic home was traditional university departments of religion or philosophy. These scholars wrote about the problems generated by the new medicine and technologies of the time, but they were not part of a discourse community that could be called an academic field or subject area. Individual scholars, working in isolation, began to legitimize bioethical issues as questions deserving rigorous academic study. But bioethics only solidified itself as a field when it became housed in institutions dedicated to the study of these questions. Academic bioethics was born with the creation of the first "bioethics center."

Ironically, academic bioethics came into existence through the creation of an institution that was not part of the traditional academy. The first institution devoted to the study of bioethical questions was a freestanding bioethics center, purposely removed from the academy with its rigid demarcations of academic study. The institution was the Hastings Center, originally called "A Center for the Study of Value and the Sciences of Man," which opened its doors in September 1970. Its founder, Daniel Callahan, created the Hastings Center to be an interdisciplinary institute solely dedicated to the serious study of bioethical questions. Callahan, a recently graduated Ph.D. in philosophy, had been one of the isolated scholars working on an issue in applied ethics, and he had found himself mired in complex questions that took him far afield from the traditional boundaries of philosophy. His topic, abortion, required engagement with the disciplines of law, medicine, and social science, which he felt himself unprepared to navigate. With academic departments functioning as islands within a university, it seemed that truly interdisciplinary work was impossible. The Hastings Center was founded to create an intellectual space for the study of these important questions from multiple perspectives and academic areas.

The second institution that helped solidify the field of bioethics was the Kennedy Institute of Ethics, which opened in 1971. The founders at Georgetown University had similar goals to those of Hastings, though they placed their center inside the traditional academy. While housed outside any particular academic departments, the Kennedy Institute came to look more like a traditional department, offering degree programs and establishing faculty appointments along a university model.

From these modest beginnings, the field of bioethics exploded, with dozens of universities following suit, creating institutions whose sole function was the study of bioethical issues. Its growth was fueled by the appearance both of new technologies, such as the artificial heart and in vitro fertilization, and new challenges, such as HIV. Bioethics was now permanently on the academic map and central to public discourse.

The Institutions of Bioethics

Whether freestanding or associated with a university, the first institutions of bioethics were centers or institutes, usually highly independent and free from the governance and promotional structure of a traditional academic department. Over the last thirty years, as the field has gained legitimacy, there has been an increasing trend of bioethics centers becoming academic departments. Originally modeled on the structure of an independent think tank, the bioethics centers of today are often housed within either a medical school or a school of arts and sciences, indistinguishable in structure from any other departments in those schools. The professionalization of bioethics has taken the field from the academic margins to the center, and with this development has come all of the trappings of traditional academics, such as tenure, degree programs, professional conferences, and academic journals.

Beginning in the 1980s, medical schools began housing bioethics institutes either as departments of medical ethics or as departments of medical humanities. Located within an undergraduate medical school, the duties of these departments include the ethics education of the M.D. students. Whereas the original bioethics centers had as their primary focus the production of scholarly research, departments of bioethics have pedagogical obligations and are viewed as institutions designed to serve the more narrow educational mission of the school. Bioethics institutions that are not housed within a medical school but instead within a school of arts and sciences have the same type of pedagogical obligations, though perhaps they serve a different student population—namely, university undergraduates or graduate students. Departments of bioethics, depending on their configuration, offer traditional undergraduate or graduate courses, undergraduate majors or concentrations, graduate degrees (usually master's degrees), undergraduate medical school ethics training, and/or residency ethics training. There are currently more than thirty master of bioethics programs in the United States, attracting a diverse student population, including recent undergraduates, students pursuing joint J.D., M.D., and Ph.D. degrees, and midcareer professionals from the fields of law, medicine, and public policy whose work requires specialty training in the field of bioethics.

Another result of the professionalization of bioethics was the pressure to publish in traditional scholarly venues, such as academic journals. But the formation of a new academic field of study necessitated the creation of academic journals in which to publish these novel scholarly works. Journals emerged that were designed solely for works in the field of bioethics, such as *The Hastings Center Report*, the *Journal of the Kennedy Institute*, the *American Journal of Bioethics*, and *Bioethics*. But the mainstreaming of bioethics into the academy also opened up space within traditional medical and scientific journals for scholarly works in bioethics. Research in bioethics is now routinely published in the likes of *JAMA* (the *Journal of the American Medical Association*), the *New England Journal of Medicine*, *Science*, and *Nature*.

Perhaps the institution most effectively used by the field of bioethics is the Internet. All major bioethics institutes, centers, or departments (and some journals) have elaborate websites, not only offering information about the specific institution, faculty, and degree programs but also undertaking an educational mission to raise the level of public debate about current bioethical issues. These websites offer substantive information for individuals seeking to become better informed about these issues. One of the most developed websites is www.bioethics.net, the companion site to the *American Journal of Bioethics*. This website not only offers actual scholarly works in the field but includes a high school bioethics project, job placement information, a "Bioethics for Beginners" section, and a daily updated collection of bioethics news stories from the popular press, with direct links to the original news articles.

The Methods of Bioethics

The founders of the field of bioethics and its first leaders were largely theologians or philosophers. Reflecting the scholarly conventions of their home disciplines, the first works in bioethics centered on a normative analysis of bioethical issues, arguing for or against the moral permissibility of a particular technology, practice, or policy. Starting in the 1970s, these philosophers and theologians were joined by physicians and lawyers, who also made normative claims about bioethical problems. But by the mid-1990s, bioethics was attracting populations of scholars who had not previously been well represented in the field—namely, social scientists and empirically trained clinicians, both physicians and nurses. With the entry of these new groups of scholars, the "methods" of bioethics began to shift, mirroring the methodologies of the new disciplines becoming central to the field. With this change, bioethics included not only normative analysis but also the empirical study of bioethical questions, what I have called "empiricized bioethics."

Empiricized bioethics takes one of two forms: either it seeks to collect empirical data needed to shed light on a bioethical problem, or it attempts to stand outside the discipline in order to study the field itself. Projects taking the first form use either qualitative or quantitative social science methodology to collect data needed to make persuasive bioethical arguments. These empirical studies might explore, for example, patient comprehension of medical information, patient and family experience with medical care, the ability of children or incompetent adults to give consent for research participation, the frequency with which practitioners face particular ethical dilemmas, and so forth.

Projects taking the second form, the study of bioethics itself, explore the way in which the field is evolving, the influence it has had on policy formation, the methods and strategies it employs, and the field's understanding of itself and its place in public life and contemporary academia. One very prominent contemporary method employing this strategy is narrative bioethics, or what might be called "deconstructionist bioethics." Using the insights of literary criticism, these bioethicists examine the discourse of the field to reveal its biases, conventions, and assumptions, making the field more self-reflective about its motives and goals. Along the same line, the field has seen the development of feminist bioethics and disability bioethics, both of which focus on issues of inclusion and exclusion, voice, and their confluence on particular substantive issues. Altogether, the empirical methods of bioethics have been so well received in the field that by the early 2000s, all bioethics centers and departments had representation from the social sciences or clinical medicine, and in many cases the empiricists constituted the majority of center or department membership.

One final methodology that has had a significant presence in medical humanities departments is literary analysis, in which literary texts are used as a vehicle for the ethics education of clinicians in training. These medical humanists use first-person illness narratives or first-person testimonies from clinicians, as well as important works in fiction, to teach health care professionals about the ethical issues involved in being both patient and practitioner.

About the Author

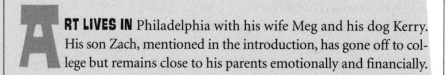

ART LIVES IN Philadelphia with his wife Meg and his dog Kerry. His son Zach, mentioned in the introduction, has gone off to college but remains close to his parents emotionally and financially.